T0180774

Springer Series in Reliability Engineering

Series editor

Hoang Pham, Piscataway, USA

Toshio Nakagawa

Random Maintenance Policies

Springer

Toshio Nakagawa
Department of Business Administration
Aichi Institute of Technology
Toyota
Japan

ISSN 1614-7839   ISSN 2196-999X   (electronic)
ISBN 978-1-4471-7181-2   ISBN 978-1-4471-6575-0   (eBook)
DOI 10.1007/978-1-4471-6575-0

Springer London Heidelberg New York Dordrecht

Printed on acid-free paper

Springer is part of Springer Science+Business Media (www.springer.com)

Preface

Reliability theory is originally to investigate the properties of randomness because most failures occur at random times. Maintenance theory is basically how to protect reliability systems from such failures by effective and practical methods. It has been well-known generally in maintenance theory that (a) preventive replacement should not be done when the failure time is exponential, and (b) optimum replacement is nonrandom [1]. This book is written mainly based on our original work for exploding the above-established theory (a) and (b) under reasonable conditions.

I have now already published four monographs *Maintenance Theory of Reliability* [2], *Shock and Damage Models in Reliability Theory* [3], and *Advanced Reliability Models and Maintenance Policies* [4] in which I have surveyed a great part of maintenance policies, using the research results of the author and my colleagues. Most of the three books have been written based on theory of stochastic processes and their mathematical tools. As the elementary text book by which a number of graduate students, researchers, and engineers can learn readily reliability theory, I have published my fourth monograph *Stochastic Processes with Applications to Reliability Theory* [5]. This was written in an easy style on stochastic processes, and all examples are quoted fittingly from reliability and maintenance models. Making good use of such writing experiences, I have now published my fifth monograph *Random Maintenance Policies* in which our recent published papers studied on random maintenance have been collected and new results are added.

The main object of this book is to write down the standard and usual maintenance theory to random maintenance theory: Some typical measures such as reliability, failure rate, and availability are defined at random times, and random age replacement and three kinds of periodic replacements are introduced in Chap. 1. New three kinds of replacement policies such as replacement first, replacement last, and replacement overtime are proposed in Chaps. 2 and 3. In particular, replacement last is a new type of replacement policies and could be expected to be used in some practical fields when a replacement cost after failure is not so high. Optimum inspection and backup policies with random checking times are derived in Chaps. 4 and 5.

Furthermore, another object of this book is (c) to form stochastically random reliability models. Parallel systems with a random number of units are proposed, and their optimization problems are discussed in Chap. 6. Using such random parallel systems, a variety of scheduling problems with random working times are solved in Chap. 7. Finally, four random reliability models are taken up, and their optimum policies are discussed in Chap. 8. In Appendix A, modified and extended types of failure rates appeared in this book are collected, and their properties are investigated theoretically. Unfortunately, there is no concrete examples of random maintenance and random systems at present, however, they would be applied certainly to actual reliability models by individual and collective efforts of practitioners and engineers in near future. In addition, several examples are cited in each chapter, and some problems given at the end of chapters with asterisk would offer a good study of research subjects to graduate students.

I wish to thank Dr. Xufeng Zhao and Dr. Satohi Mizutani for all chapters, Dr. Kenichiro Naruse for Chap. 5, Dr. Kodo Ito for Chap. 6 and Professor Mingchih Chen for Chap. 7, who are co-workers on our research papers. Especially, I wish to express my special thanks to Dr. X. Zhao and Dr. S. Mizutani in writing and typing this book and for their careful reviews. Finally, I would like to express my sincere appreciation to Professor Hoang Pham, Rutgers University, and Editor Anthony Doyle, Springer-Verlag, London, for providing me the opportunity to write this book.

Toyota, May 2013   Toshio Nakagawa

References

1. Barlow RE, Proschan F (1965) Mathematical theory of reliability. Wiley, New York
2. Nakagawa T (2005) Maintenance theory of reliability. Springer, London
3. Nakagawa T (2007) Shock and damage models in reliability theory. Springer, London
4. Nakagawa T (2008) Advanced reliability models and maintenance policies. Springer, London
5. Nakagawa T (2011) Stochastic processes with applications to reliability theory. Springer, London

Contents

Chapter 1
Introduction

Random models frequently appear in probability [1] and in statistics [2], and are applied in a variety of fields such as agriculture, biology, animal breeding, applied genetics, econometrics, quality control, medicine, engineering, education, and environmental and social sciences [2]. The close connections of *randomness* to other scientific fields such as computability and complexity theory, information theory, biology, physics, quantum mechanics, learning theory, and artificial intelligence are largely collected [3]. In particular, the relationships between randomness and computability are completely surveyed [4, 5].

Similarly, reliability theory was also originally to investigate the properties of failures occurring at random times, and maintenance theory was basically to study how to protect from such random failures by getting exact and full information about them. The most typical continuous distribution of representing completely *at random* would be an exponential distribution, which has the memoryless property [1, p. 243], [6, p. 74], [7, p. 13] and whose failure rate is constant. It has been well known that the distribution of the time between failures tends to be exponential theoretically and practically as objective systems become more complex and large scale, and consist of much more different units [8, p. 18].

It has been already established in reliability theory that:

(a) When the failure time is exponential, an operating unit should be replaced only at failure.
(b) An optimum age replacement is nonrandom [8, p. 86] (Problem 1 in Sect. 1.4).

It would be indispensably necessary to take a certain maintenance for more complex systems, even if their failure times might be exponential. Furthermore, as systems become much more random, random maintenance fitting for them should be done rather than a deterministic one. It was shown [8, p. 72] that when systems work in a variable cycle, the maintenance policy would have to be random taking advantage of any free time available for maintenance.

The book is written based on our original work mainly for answering the above questions: Some typical measures such as reliability, failure rate, and availability are defined newly at random time, and the expected number of failures until ran-

© Springer-Verlag London 2014
T. Nakagawa, *Random Maintenance Policies*,
Springer Series in Reliability Engineering, DOI 10.1007/978-1-4471-6575-0_1

dom time is obtained in Chap. 1. In addition, when failure distributions are uncertain, replacement policies are proposed, using a uniform distribution and the entropy model. Furthermore, a random age replacement and three kinds of periodic replacements are introduced, and their expected cost rates are obtained and optimum policies are derived.

For answering question (a), we introduce the shortage and excess costs for a random age replacement and show that a finite replacement time exists even for exponential failure times in Chap. 2. Furthermore, three kinds of age and periodic replacement policies with constant time T and random working time Y are proposed in Chaps. 2 and 3 [9]:

(1) The unit is replaced at time T or at time Y, whichever occurs first.
(2) The unit is replaced at time T or at time Y, whichever occurs last.
(3) The unit is replaced at the first completion of working times over time T.

They are called *replacement first*, *replacement last*, and *replacement overtime*, respectively. The expected cost rates of each policy are obtained, and optimum policies which minimize them are derived analytically. Replacement last is a new theoretical scheme of replacement policies. If the replacement cost after failure was estimated to be not so high, then the unit should be used as long as possible before its failure. We compare the standard policy with only replacement time T to the above three policies and discuss analytically and numerically which policy is better.

It has been well known in periodic inspection policies that a finite inspection time exists even if failure times are exponential [10, p. 203]. We apply the notion of random replacement to inspection policies in Chap. 4: We introduce a random inspection policy in which the unit is checked at random working times Y_j ($j = 1, 2, \ldots$). In addition, when the unit is also checked at periodic and sequential times, the total expected costs until failure detection, and optimum policies which minimize them are derived. Furthermore, by similar methods of Chaps. 2 and 3, three kinds of policies of *inspection first*, *inspection last*, and *inspection overtime* are newly proposed, their optimum policies are discussed and compared analytically and numerically with each other, when failure times are exponential. A random inspection policy for a finite interval is also considered. In Chap. 5, we apply the random inspection policy to the backup system in which the backup operation is executed to the latest checkpoint when failures are detected [11, p. 123]: Optimum periodic and random backup policies which minimize the total expected cost until backup operation are derived. Furthermore, several random checkpoint models are presented and their optimum policies are discussed.

Another proposal of the book is:

(c) To form stochastically *random reliability systems* and analyze them.

At the beginning of such essential researches, we introduce a parallel redundant system with a random number of units in Chap. 6 [12]. The mean time to failure is obtained. An optimum number of units and replacement time for the system which minimize the expected cost rates are derived analytically and numerically. Next, we take up a K-out-of-n system in which K is a random variable and derive an optimum number n^* when K is a geometric and Poisson distributions (Problem 2 in Sect. 1.4).

Scheduling problems of jobs often appear in production management [13]. Chapter 7 surveys the scheduling models with a random working time [11, p. 82], [14]: Next, we consider the scheduling time for N works in tandem and in parallel, and derive their optimum schedules. Furthermore, we solve another problem in which how many number of units and what kinds of redundant systems we have to provide for two kinds of works. As one of extended models, we consider the case where N is a random variable and rewrite afresh all results derived in this chapter.

In addition, we take up the following four random models in Chap. 8 and discuss optimum policies for each model (Problem 3 in Sect. 1.4):

(1) Inspection, replacement, and preventive maintenance for a random finite interval.
(2) Random interval reliability and its preventive maintenance policy.
(3) Cumulative damage models with random failure level and their replacement policies.
(4) Other reliability models with random scheduling time, random number of inspections and failures.

In Appendix A, extended failure rates that appear in the book are collected and their properties are discussed theoretically when the failure rate increases and working times are exponential. Furthermore, some interesting inequalities among failure rates are shown. These results would be more useful for analyzing other maintenance policies with such failure rates.

1.1 Further Studies and Applications

The book attains three main purposes to study on: (a) Maintenance policies for exponential failure distributions, (b) optimum random maintenance policies for various reliability models, and (c) stochastic modeling of random systems. We cannot get good answers for (a) only except introducing shortage and excess costs in Sect. 2.1.1. Recently, systems become much more complex as maintenance and reliability techniques have been developed and expanded, and so that, their failure times would tend to be exponential. For further studies, it would be necessary to consider appropriate maintenance counterplans by any means for much more random and complex systems with types of exponential failure times.

It has been shown in (b) [8, p. 72] that systems with random working and processing times might be suitably maintained at random times because it would be impractical to do some maintenance in strictly periodic times. After that, however, it seems that there is no discussion and no application of random maintenance policies in reliability theory. Good results of random maintenance policies for age and periodic replacements in Chaps. 2 and 3, and inspection in Chap. 4 would offer new interesting topics to theoretical researchers. Especially, *maintenance last* is a new general idea in maintenance theory. This would be brought in many reliability systems and studied more theoretically. Unfortunately, there exists now no concrete example of random maintenance at present. However, aged plants and old public infrastructures have increased remarkably in advanced nations, and the importance of their maintenance would be much higher from the viewpoints of safety, economics, and circumstances. Modifying and extending these results of random maintenance, they would be applied certainly to actual systems by practical engineers.

From the above discussion, we know that systems will become much more complex, and old plants and structures will increase rapidly in the near future. In simple words, some systems tend to become random more and more because their structures become vague. As time goes by, system actualities will also become less certain. A parallel system with random number of units as an example of aircraft fuselage [12] proposed in Chap. 6 is the first model to study random systems in the future. A K-out-of-n system when K is a random variable might be more realistic. In fact, when n is large, it would be impossible to determine a specific number K, however, it would be possible to estimate K statistically. Unfortunately, there is no actual good example of such systems. We believe that more random systems would be schemed out, studied theoretically, and applied to actual systems practically. Some random models in Chap. 8 are just extended examples of randomizing reliability models. Such techniques would give new effective tools for analyzing other reliability models.

Furthermore, problems with asterisks at the end of each chapter would offer suggestive subjects to researchers such as graduate students by making suitable modifications and applications.

1.2 Random Failure Time

1.2.1 Reliability, Failure Rate, and Availability

The most concern in reliability theory is to know the probabilistic and statistical properties of *reliability* $R(t)$, *failure rate* $h(t)$, and *availability* $A(t)$ at a given point of time t ($0 \le t < \infty$). Such reliability measures are roughly defined as follows: When the unit has a failure distribution $F(t)$ and a density function $f(t) \equiv dF(t)/dt$, $R(t) \equiv 1 - F(t)$ is the probability that it has continued to operate without failure until time t, failure rate is $h(t) \equiv f(t)/R(t)$ and $h(t)\Delta t$ means the probability that with age t it will fail in $(t, t + \Delta t]$, and $A(t)$ is the probability that it will operate at time t. These measures of reliability models from large-scale systems to small parts have been investigated theoretically and estimated statistically to analyze them and to apply practically them to maintenance policies.

We sometimes encounter situations when working and processing times required for jobs are needed and severe events suddenly happen at time T which might occur at random times. In such situations, we have to investigate reliability measures at a random time T which would be distributed according to a general distribution $G(t) \equiv \Pr\{T \le t\}$ for $t \ge 0$ with finite mean $1/\theta$ ($0 < \theta < \infty$). Then, the above measures are newly defined: Random reliability is

$$\widetilde{R}(G) \equiv \int_0^\infty R(t)\,dG(t), \tag{1.1}$$

random failure rate is

$$\widetilde{h}(G) \equiv \int_0^\infty h(t)\,dG(t), \tag{1.2}$$

and random availability is

$$\widetilde{A}(G) \equiv \int_0^\infty A(t)\,dG(t). \tag{1.3}$$

Example 1.1 (Reliability for exponential random time) When $G(t) = 1 - e^{-\theta t}$ ($0 < \theta < \infty$) and $F(t) = 1 - e^{-\lambda t}$ ($0 < \lambda < \infty$), i.e., $R(t) = e^{-\lambda t}$, from (1.1),

$$\widetilde{R}(\theta) = \frac{\theta}{\theta + \lambda},$$

which decreases with $1/\theta$ from 1 to 0 (Problem 4 in Sect. 1.4). When the mean random time $E\{T\} = 1/\theta$ is equal to t, i.e., $t = 1/\theta$, we compare $R(t) = e^{-\lambda t}$ and $\widetilde{R}(1/t) = 1/(1 + \lambda t)$. Clearly, $\widetilde{R}(1/t) > R(t)$ because $e^{\lambda t} > 1 + \lambda t$ for $t > 0$. In this case, random reliability is greater than reliability $R(t)$.

Next, when the failure time has a Weibull distribution $F(t) = 1 - \exp(-\lambda t^\alpha)$, i.e., $h(t) = \lambda \alpha t^{\alpha-1}$ ($\alpha > 0$), from (1.2), random failure rate is

$$\widetilde{h}(\theta) = \frac{\lambda \alpha \Gamma(\alpha)}{\theta^{\alpha-1}}, \tag{1.4}$$

where $\Gamma(\alpha) \equiv \int_0^\infty x^{\alpha-1} e^{-x} dx$ ($\alpha > 0$). Random failure rate decreases with $1/\theta$ for $\alpha < 1$, is constant λ for $\alpha = 1$, and increases for $\alpha > 1$, which is the same property as $h(t)$. In addition, when $t = 1/\theta$, $\widetilde{h}(\theta) < h(t)$ for $1 < \alpha < 2$ and $\widetilde{h}(\theta) \geq h(t)$ for the others (Problem 5 in Sect. 1.4). □

Example 1.2 (Availability for one-unit system with repair) Consider a standard one-unit system which repeats the operating and repair states alternately [8, p. 76], [10, p. 40]. It is assumed that the failure time of an operating unit has a general distribution $F_O(t)$ with finite mean $1/\lambda$ and the repair time of a failed unit has a general distribution $F_R(t)$ with finite mean $1/\mu$ ($0 < \mu < \infty$). Then, the Laplace-Stieltjes transform (LS) of the probability $A(t)$ that the unit is operating at time t, given that it begins to operate at time 0, is

$$A^*(s) \equiv \int_0^\infty e^{-st} dA(t) = \frac{1 - F_O^*(s)}{1 - F_O^*(s) F_R^*(s)}, \tag{1.5}$$

where the asterisk of the function denotes the LS transform with itself, i.e., $\Phi^*(s) \equiv \int_0^\infty e^{-st} d\Phi(t)$ for $Re(s) > 0$ [6, p. 241], [7, p. 219]. Thus, when $G(t) = 1 - e^{-\theta t}$, random availability is, from (1.3) and (1.5),

$$\tilde{A}(\theta) = \int_0^\infty A(t)\theta e^{-\theta t} \, dt = \int_0^\infty e^{-\theta t} \, dA(t) = A^*(\theta). \tag{1.6}$$

In particular, when $F_O(t) = 1 - e^{-\lambda t}$ and $F_R(t) = 1 - e^{-\mu t}$ [8, p. 77], [10, p. 43],

$$A(t) = \frac{\mu}{\lambda + \mu} + \frac{\lambda}{\lambda + \mu} e^{-(\lambda+\mu)t}, \tag{1.7}$$

$$\tilde{A}(\theta) = \frac{\theta + \mu}{\theta + \lambda + \mu}. \tag{1.8}$$

Both availabilities decrease strictly with t and $1/\theta$ from 1 to $\mu/(\lambda + \mu)$, which is called limiting availability. When $t = 1/\theta$, $A(t) < \tilde{A}(\theta)$ for $0 < t < \infty$, because

$$\frac{\mu}{\lambda + \mu} + \frac{\lambda}{\lambda + \mu} e^{-(\lambda+\mu)t} < \frac{1 + \mu t}{1 + (\lambda + \mu)t}.$$

This means that random availability is greater than availability at time t when $t = 1/\theta$ (Problem 6 in Sect. 1.4). $\qquad\square$

1.2.2 Expected Number of Failures

We are interested in the expected number of failures $M(t)$ during $(0, t]$ when units are replaced at each failure, and $H(t)$ when units undergo only minimal repair at each failure [8, p. 96], [10, p. 96]. The expected number $M(t)$ is called the *renewal function* in a renewal process of stochastic processes and $H(t)$ is called the *mean value function* in a nonhomogeneous Poisson process [6, p. 77], [7, p. 61].

When the unit fails according to a general distribution $F(t)$ and is replaced immediately, the renewal function is

$$M(t) = \sum_{j=1}^\infty F^{(j)}(t), \tag{1.9}$$

where $\Phi^{(j)}(t)$ denotes the j-fold Stieltjes convolution of $\Phi(t)$ with itself, and $\Phi^{(0)}(t) \equiv 1$ for $t \geq 0$. When the unit undergoes minimal repair at failures and operates again immediately,

$$F(t) = 1 - e^{-H(t)}, \quad \text{i.e.,} \quad H(t) = -\log R(t), \tag{1.10}$$

where note that $H(t)$ is also called the *cumulative hazard rate* in reliability theory because $H(t) \equiv \int_0^t h(u)\,du$. The properties of functions $M(t)$ and $H(t)$ were obtained and compared with each other [7, p. 61]. For example, when $F(t)$ is IFR, i.e., $h(t)$ increases, for $0 < t < \infty$,

$$\frac{H(t)}{1 + H(t)} \leq F(t) \leq M(t) \leq H(t) \leq \frac{F(t)}{\overline{F}(t)},$$

$$\frac{M(t)}{t} \leq \frac{F(t)}{\int_0^t \overline{F}(u)\,du} \leq \frac{H(t)}{t} \leq h(t) \leq \frac{\overline{F}(t)}{\int_t^\infty \overline{F}(u)\,du}.$$

More relationships among extended failure rates are shown in Appendix A.

By the similar way of defining random reliability, the random renewal function is

$$\tilde{M}(G) = \int_0^\infty M(t)\,dG(t), \tag{1.11}$$

and the random cumulative hazard rate is

$$\tilde{H}(G) \equiv \int_0^\infty H(t)\,dG(t). \tag{1.12}$$

Both $\tilde{M}(G)$ and $\tilde{H}(G)$ represent the expected number of failures during $[0, T]$ in which T is a random variable with a general distribution $G(t)$.

For a one-unit system with a failure distribution $F(t)$, from (1.9),

$$\tilde{M}(G) = \sum_{j=1}^\infty \int_0^\infty F^{(j)}(t)\,dG(t), \tag{1.13}$$

and from (1.10),

$$\tilde{H}(G) = -\int_0^\infty \log R(t)\,dG(t). \tag{1.14}$$

Example 1.3 *(Renewal function for exponential random time)* When $G(t) = 1 - \mathrm{e}^{-\theta t}$, (1.13) is

$$\widetilde{M}(\theta) = \sum_{j=1}^{\infty} \int_0^{\infty} F^{(j)}(t)\theta \mathrm{e}^{-\theta t}\, dt = \sum_{j=1}^{\infty} [F^*(\theta)]^j = \frac{F^*(\theta)}{1 - F^*(\theta)}. \qquad (1.15)$$

In addition, when $F(t) = 1 - \mathrm{e}^{-\lambda t}$, $M(t) = \lambda t$, and $F^*(\theta) = \lambda/(\theta + \lambda)$, and hence, $\widetilde{M}(\theta) = \lambda/\theta$, which is equal to $M(t) = \lambda t$ when $t = 1/\theta$.

Next, suppose that $F(t)$ is a gamma distribution, i.e., $F(t) = 1 - \sum_{j=0}^{k-1} [(\lambda t)^j / j!] \mathrm{e}^{-\lambda t}$ $(k = 1, 2, \ldots)$ and $F^*(\theta) = [\lambda/(\theta + \lambda)]^k$. Then, the renewal function is [7, p. 52]

$$M(t) = \sum_{n=1}^{\infty} \sum_{j=nk}^{\infty} \frac{(\lambda t)^j}{j!} \mathrm{e}^{-\lambda t}, \qquad (1.16)$$

and

$$\lim_{t \to \infty} \frac{M(t)}{t} = \frac{\lambda}{k}.$$

On the other hand, the random renewal function is, from (1.15),

$$\widetilde{M}(\theta) = \frac{\lambda^k}{(\theta + \lambda)^k - \lambda^k}, \qquad (1.17)$$

and

$$\lim_{\theta \to 0} \theta \widetilde{M}(\theta) = \frac{\lambda}{k}.$$

It can be easily seen that $\widetilde{M}(\theta) \le \lambda/(k\theta)$.

Table 1.1 presents $M(t)$, $\widetilde{M}(\theta)$, and t/k when $t = 1/\theta$, $F(t) = 1 - \sum_{j=0}^{k-1} (t^j / j!)\mathrm{e}^{-t}$ $(k = 1, 2, 3)$, and $G(t) = 1 - \mathrm{e}^{-\theta t}$ (Problem 7 in Sect. 1.4). This indicates that $M(t) \le \widetilde{M}(\theta) \le t/k$, and as t becomes larger, $\widetilde{M}(\theta)$ approaches to $M(t)$ very well. As a result, $\widetilde{M}(\theta)$ can be much easily computed and gives a good upper bound of $M(t)$ for large t.

When $F(t)$ is a Weibull distribution, i.e., $F(t) = 1 - \exp(-\lambda t^{\alpha})$ $(\alpha > 0)$ and $H(t) = \lambda t^{\alpha}$,

$$\widetilde{H}(\theta) = \int_0^{\infty} \lambda t^{\alpha} \theta \mathrm{e}^{-\theta t}\, dt = \frac{\lambda \Gamma(\alpha + 1)}{\theta^{\alpha}} = \frac{\widetilde{h}(\theta)}{\theta}.$$

Table 1.1 Comparisons of $M(t)$, $\tilde{M}(\theta)$, and t/k when $t = 1/\theta$ and $F(t) = 1 - \sum_{j=0}^{k-1}(t^j/j!)e^{-t}$

t	$k = 1$	$k = 2$			$k = 3$		
	$M(t) = \tilde{M}(\theta)$	$M(t)$	$\tilde{M}(\theta)$	$t/2$	$M(t)$	$\tilde{M}(\theta)$	$t/3$
1	1	0.00	0.01	0.5	0.00	0.00	0.3
2	2	0.02	0.04	1.0	0.00	0.00	0.7
5	5	0.12	0.23	2.5	0.00	0.01	0.17
10	10	0.47	0.83	5.0	0.02	0.08	3.3
20	20	1.76	2.86	10.0	0.11	0.47	6.7
50	50	9.20	12.50	25.0	1.44	3.85	16.7
100	100	28.38	33.33	50.0	8.09	14.29	33.3
200	200	75.46	80.00	100.0	34.01	42.11	66.7
500	500	225.00	227.27	250.0	133.32	137.36	166.7
1,000	1,000	475.00	476.19	500.0	300.00	302.11	333.3

When $t = 1/\theta$, $\tilde{H}(\theta) < H(t)$ for $0 < \alpha < 1$, $\tilde{H}(\theta) = H(t)$ for $\alpha = 1$, and $\tilde{H}(\theta) > H(t)$ for $\alpha > 1$. In general, because $H(t) = \int_0^t h(u)du$, $\tilde{H}(\theta) = \tilde{h}(\theta)/\theta$ from the property of LS transform [7, p. 222] (Problem 8 in Sect. 1.4). □

Example 1.4 (Renewal function for one-unit system with repair) Consider the same one-unit system in Example 1.2. When the unit begins to operate at time 0, the expected number $M(t)$ of failures during $[0, t]$ is

$$M^*(s) \equiv \int_0^\infty e^{-st} \, dM(t) = \frac{F_O^*(s)}{1 - F_O^*(s)F_R^*(s)}. \tag{1.18}$$

When $G(t) = 1 - e^{-\theta t}$,

$$\tilde{M}(\theta) = \int_0^\infty M(t)\theta e^{-\theta t} \, dt = M^*(\theta).$$

In particular, when $F_O(t) = 1 - e^{-\lambda t}$ and $F_R(t) = 1 - e^{-\mu t}$ [8, p. 76], [10, p. 43],

$$M(t) = \frac{\lambda \mu t}{\lambda + \mu} + \left(\frac{\lambda}{\lambda + \mu}\right)^2 [1 - e^{-(\lambda+\mu)t}],$$

$$\tilde{M}(\theta) = \frac{\lambda(\theta + \mu)}{\theta(\theta + \lambda + \mu)},$$

and hence, $\tilde{M}(\theta) < M(t)$ when $t = 1/\theta$ for $0 < t < \infty$ (Problem 9 in Sect. 1.4). In exponential cases, when $t = 1/\theta$, $\tilde{A}(\theta) > A(t)$ and $\tilde{M}(\theta) < M(t)$. □

1.2.3 Uniform Distribution

A uniform distribution is well known as a failure distribution in the case where failures occur only at a Poisson process for a finite interval and when a failure was detected at time t, without any information of its failure time [6, p. 71], [7, p. 24]. This uniform distribution also has appeared in an inspection model for a finite interval [7, p. 24], [8, p. 113], [11, p. 96] and a backup model [11, p. 89].

 We consider an operating unit in which its failure has certainly occurred in $[0, S]$ $(0 < S < \infty)$, without any information of its occurrence time, i.e., a failure occurs randomly in $[0, S]$. We assume that the failure time is uniformly distributed over $[0, S]$, and as the preventive replacement, we will apply it to make an age replacement policy with replacement time T $(0 < T \leq S)$ in Chap. 2: Let c_T be the replacement cost at time T and c_F be the replacement cost at failure with $c_F > c_T$. Because the mean time to failure is

$$\int_0^T \frac{t}{S}\, dt + \left(1 - \frac{T}{S}\right) T = \frac{T(2S - T)}{2S},$$

the expected cost rate is, from (2.1),

$$C(T; S) = \frac{c_F T + c_T (S - T)}{T(S - T/2)}. \tag{1.19}$$

Clearly,

$$C(0; S) \equiv \lim_{T \to 0} C(T; S) = \infty, \qquad C(S; S) \equiv \lim_{T \to S} C(T; S) = \frac{2c_F}{S}.$$

Differentiating $C(T; S)$ with respect to T and setting it equal to zero,

$$\frac{T^2}{2S(S - T)} = K,$$

i.e.,

$$\left(\frac{T}{S}\right)^2 + 2K \left(\frac{T}{S}\right) - 2K = 0, \tag{1.20}$$

where

$$K \equiv \frac{c_T}{c_F - c_T}.$$

Solving (1.20) for T/S, an optimum T^* is given by

$$\frac{T^*}{S} = \sqrt{K^2 + 2K} - K < 1.$$

Clearly, T^*/S increases with K from 0 to 1, i.e., increases with c_T/c_F. In particular, when $K = 1/2$, i.e., $c_T/c_F = 1/3$, $T^*/S = (\sqrt{5} - 1)/2 \approx 0.618$ which is equal to the golden ratio [11, p. 19, p. 200] (Problem 10 in Sect. 1.4). Thus, we can easily compute an optimum ratio of T^* to S for given c_T/c_F.

1.2.4 Entropy Model

The entropy model [11, p. 199], [17] used in marketing sciences was applied to an age replacement policy as follows: It is assumed that the price of brand $A(B)$ is $c_1(c_2)$, respectively. A consumer buys $A(B)$ with selection rate $p(q)$, respectively, where $p + q = 1$. Then, the mean purchase cost is $C \equiv c_1 p + c_2 q$ and the entropy is $H \equiv -p \log p - q \log q$. We want to minimize C and maximize H, i.e., we maximize an objective function

$$C(p, q) \equiv \frac{H}{C} = \frac{-p \log p - q \log q}{c_1 p + c_2 q}, \tag{1.21}$$

subject to $p + q = 1$.

We apply the above entropy model to the age replacement with a planned time T ($0 < T \le \infty$): By replacing $p = F(T), q = \overline{F}(T)$, and $c_1 = c_F, c_2 = c_T$, (1.21) is rewritten as

$$C(T) = \frac{-F(T) \log F(T) - \overline{F}(T) \log \overline{F}(T)}{c_F F(T) + c_T \overline{F}(T)}. \tag{1.22}$$

We find an optimum T^* which maximizes $C(T)$. Differentiating $C(T)$ with respect to T and setting it equal to zero,

$$\frac{\log F(T)}{\log \overline{F}(T)} = \frac{c_F}{c_T}. \tag{1.23}$$

Example 1.5 (Uniform distribution and entropy model) Table 1.2 presents T^*/S in (1.20), $F(T)$ in (1.23), and $F(T^*)$ in Table 9.8 of [11, p. 202] when $F(t) = 1 - \exp[-(\lambda t)^\alpha]$ (Problem 11 in Sect. 1.4). This indicates that T^*/S has good approximations to $F(T^*)$ for small c_F/c_T, and conversely, $F(T)$ has good ones for large c_F/c_T. The values of T^*/S and $F(T)$ would be useful for rough approximations of optimum $F(T^*)$, because they can be estimated and computed easily, irrespective of a failure distribution. □

Table 1.2 T^*/S in (1.20), $F(T)$ in (1.23), and $F(T^*)$ in Table 9.8 of [11, p. 202]

c_F/c_T	$T^*/S \times 100$	$F(T) \times 100$	$F(T^*) \times 100$			
			$\alpha = 1.6$	$\alpha = 2.0$	$\alpha = 2.4$	$\alpha = 3.0$
2	73.2	38.2	91	70	55	41
4	54.9	27.6	46	30	22	16
6	46.3	22.2	30	19	14	10
10	37.2	16.5	18	11	8	5

1.3 Random Replacement

It is assumed that an operating unit has a failure time X with a general distribution $F(t) \equiv \Pr\{X \leq t\}$, a density function $f(t) \equiv \mathrm{d}F(t)/\mathrm{d}t$, a failure rate $h(t) \equiv f(t)/\overline{F}(t)$, a cumulative hazard rate $H(t) \equiv \int_0^t h(u)\mathrm{d}u$, a renewal function $M(t) \equiv \sum_{j=1}^{\infty} F^{(j)}(t)$ and $m(t) \equiv \mathrm{d}M(t)/\mathrm{d}t$, where $\overline{F}(t) = 1 - F(t)$ and $F^{(j)}(t)$ $(j = 1, 2, \ldots)$ is the j-fold Stieltjes convolution of $F(t)$ with itself, and $F^{(0)}(t) \equiv 1$ for $t \geq 0$. We consider the following four random replacement policies: The unit is replaced at a random time Y which has a general distribution $G(t) \equiv \Pr\{Y \leq t\}$ with finite mean $1/\theta$ $(0 < \theta < \infty)$.

1.3.1 Random Age Replacement

The unit is replaced at time Y or at failure, whichever occurs first. Then, the probability that the unit is replaced at time Y is

$$\Pr\{Y \leq X\} = \int_0^{\infty} \overline{F}(t)\,\mathrm{d}G(t),$$

and the probability that it is replaced at failure is

$$\Pr\{X \leq Y\} = \int_0^{\infty} \overline{G}(t)\,\mathrm{d}F(t).$$

Thus, the mean time to replacement is

$$\int_0^{\infty} t\overline{F}(t)\,\mathrm{d}G(t) + \int_0^{\infty} t\overline{G}(t)\,\mathrm{d}F(t) = \int_0^{\infty} \overline{G}(t)\overline{F}(t)\,\mathrm{d}t.$$

Therefore, the expected cost rate is, from [8, p. 86],

$$C_A(G) = \frac{c_F \int_0^\infty \overline{G}(t)\,dF(t) + c_R \int_0^\infty \overline{F}(t)\,dG(t)}{\int_0^\infty \overline{G}(t)\overline{F}(t)\,dt}, \tag{1.24}$$

where c_R = replacement cost at random time Y and c_F = replacement cost at failure with $c_F > c_R$. In particular, when $G(t) = 1 - e^{-\theta t}$, the expected cost rate in (1.24) is a function of θ and is given by

$$C_A(\theta) = \frac{c_R + (c_F - c_R)F^*(\theta)}{[1 - F^*(\theta)]/\theta}, \tag{1.25}$$

where $\Phi^*(s)$ is the LS transform of any function $\Phi(t)$. In a similar way, when $F(t) = 1 - e^{-\lambda t}$,

$$C_A(G) = \frac{c_F - (c_F - c_R)G^*(\lambda)}{[1 - G^*(\lambda)]/\lambda}. \tag{1.26}$$

Example 1.6 (*Random replacement for gamma failure time*) When $G(t) = 1 - e^{-\theta t}$ and $F(t) = 1 - (1 + \lambda t)e^{-\lambda t}$, the expected cost rate in (1.25) is

$$C_A(\theta) = \frac{c_R + (c_F - c_R)[\lambda/(\theta + \lambda)]^2}{\{1 - [\lambda/(\theta + \lambda)]^2\}/\theta} = \frac{c_R(\theta + \lambda)^2 + (c_F - c_R)\lambda^2}{\theta + 2\lambda}.$$

Differentiating $C_A(\theta)$ with respect to θ and setting it equal to zero,

$$\left(\frac{\theta}{\lambda} + 2\right)^2 = \frac{c_F}{c_R}.$$

Thus, if $c_F/c_R \leq 4$, then $\theta^* \to 0$ and $C_R(0) = c_F\lambda/2$. In this case, we should make no random replacement. Conversely, if $c_F/c_R > 4$, then $\theta^* = \lambda(\sqrt{c_F/c_R} - 2)$ (Problem 12 in Sect. 1.4). $\qquad\square$

1.3.2 Random Periodic Replacement

The unit is replaced at periodic times kY ($k = 1, 2, \ldots$) and undergoes minimal repair at each failure between replacements. Then, because failures occur at a non-homogeneous Poisson process with mean value function $H(t)$ [7, p. 27], [10, p. 102], the expected number of failures between replacements is

$$\int_0^\infty H(t)\,dG(t) = \int_0^\infty \overline{G}(t)h(t)\,dt,$$

and the mean time to replacement is $1/\theta$. Thus, the expected cost rate is

$$C_P(G) = \theta \left[c_M \int_0^\infty \overline{G}(t) h(t)\, dt + c_R \right], \tag{1.27}$$

where c_M = minimal repair cost at each failure and c_R is given in (1.24).

Example 1.7 (Periodic replacement for Weibull failure time) When $G(t) = 1 - e^{-\theta t}$ and $H(t) = \lambda t^\alpha$ $(\alpha > 1)$, the expected cost rate in (1.27) is

$$C_P(\theta) = \frac{c_M \lambda \Gamma(\alpha+1)}{\theta^{\alpha-1}} + c_R \theta.$$

An optimum θ^* which minimizes $C_P(\theta)$ is easily given by

$$\frac{1}{\theta^*} = \left[\frac{c_R}{c_M \lambda(\alpha-1)\Gamma(\alpha+1)} \right]^{1/\alpha}.$$

□

1.3.3 Random Block Replacement

The unit is replaced at periodic times kY $(k = 1, 2, \dots)$ and also is replaced with a new one at each failure between replacements. Because the expected number of failures between replacements is

$$\int_0^\infty M(t)\, dG(t) = \int_0^\infty \overline{G}(t) m(t)\, dt.$$

Thus, the expected cost rate is

$$C_B(G) = \theta \left[c_F \int_0^\infty \overline{G}(t) m(t)\, dt + c_R \right], \tag{1.28}$$

where c_F = replacement cost at each failure and c_R is given in (1.24). In particular, when $G(t) = 1 - e^{-\theta t}$,

$$C_B(\theta) = \theta \left[\frac{c_F F^*(\theta)}{1 - F^*(\theta)} + c_R \right].$$

Example 1.8 (Block replacement for gamma failure time) When $F(t) = 1 - (1 + \lambda t)e^{-\lambda t}$, the expected cost rate is

$$C_B(\theta) = \frac{c_F \lambda^2}{\theta + 2\lambda} + c_R \theta.$$

An optimum θ^* which minimizes $C_B(\theta)$ is

$$\left(\frac{\theta}{\lambda} + 2\right)^2 = \frac{c_F}{c_R},$$

which agrees with Example 1.6 (Problem 12 in Sect. 1.4). ☐

1.3.4 No Replacement at Failure

The unit is replaced at periodic times kY ($k = 1, 2 \dots$), however, it is not replaced at failure and remains in a failed state for the time interval from a failure to its replacement. Then, the mean time from failure to replacement is

$$\int_0^\infty \left[\int_0^t (t - u) \, dF(u) \right] dG(t) = \int_0^\infty \overline{G}(t) F(t) \, dt.$$

Thus, the expected cost rate is

$$C_D(G) = \theta \left[c_D \int_0^\infty \overline{G}(t) F(t) \, dt + c_R \right], \tag{1.29}$$

where c_D = downtime cost per unit of time from failure to replacement and c_R is given in (1.24). In particular, when $G(t) = 1 - e^{-\theta t}$,

$$C_D(\theta) = c_D F^*(\theta) + c_R \theta.$$

Example 1.9 (Replacement for gamma failure time) When $f(t) = [\lambda(\lambda t)^{\alpha-1}/\Gamma(\alpha)]e^{-\lambda t}$ ($\alpha > 0$), the expected cost rate is

$$C_D(\theta) = c_D \left(\frac{\lambda}{\theta + \lambda}\right)^\alpha + c_R \theta.$$

An optimum θ^* which minimizes $C_D(\theta)$ is given by

$$\alpha \left(\frac{\lambda}{\theta + \lambda} \right)^{\alpha+1} = \frac{c_R \lambda}{c_D}.$$

If $\alpha \le c_R \lambda / c_D$, then $\theta^* \to 0$, i.e., we should make no random replacement and the unit always remains in a failed state.   □

In general, we can summarize the expected cost rates of random periodic replacements in Sects. 1.3.2, 1.3.3 and 1.3.4 as follows:

$$C(G) = \theta \left[c_i \int_0^\infty \overline{G}(t) \varphi(t)\, dt + c_R \right], \qquad (1.30)$$

where $i =$ M, F, D and $\varphi(t) = h(t), m(t), F(t)$, respectively.

1.4 Problems

1. Make certain that (a) and (b) are true in reliability theory [8, p. 86], [10, p. 74].
2. Consider other random systems in which some system parameters are random and analyze them.
3. Consider other random models in which some maintenance parameters are random and analyze them.
4. Make certain that when $\lambda = 1$ and $1/\theta = (\sqrt{5} - 1)/2$ which is the golden ratio, $\widetilde{R}(\theta) = 1/\theta = (\sqrt{5} - 1)/2$.
5. Prove that $\widetilde{h}(\theta) < h(t)$ for $1 < \alpha < 2$ and $\widetilde{h}(\theta) \ge h(t)$ for the others.
6. Prove that for $0 < t < \infty$,

$$\frac{\mu}{\lambda + \mu} < \frac{\mu}{\lambda + \mu} + \frac{\lambda}{\lambda + \mu} e^{-(\lambda + \mu)t} < \frac{1 + \mu t}{1 + (\lambda + \mu)t}.$$

7. Compute $M(t)$, $\widetilde{M}(\theta)$ and t/k when $k = 4$, and prove that when $t = 1/\theta$ and $k = 2$, $\lambda t/2 > \widetilde{M}(1/t) > M(t)$ for $0 < t < \infty$, i.e.,

$$\frac{\lambda t}{2} - \frac{1}{4}(1 - e^{-2\lambda t}) < \frac{\lambda^2 t^2}{1 + 2\lambda t} < \frac{\lambda t}{2}.$$

8. Make certain that when $F(t) = \int_0^t f(u)\, du$, the LS transform of $F(t)$ is $F^*(s) = \int_0^\infty e^{-st}\, dF(t) = f^*(s)/s$ [7, p. 222].
9. Prove that $\widetilde{M}(\theta) < M(t)$ when $t = 1/\theta$ for $0 < t < \infty$.
10. Investigate other examples of the golden ratio appeared in reliability models.
11. Compute numerically T^*/S, $F(T)$ and $F(T^*)$ in other cases and compare them.
12. Compute θ^* numerically in Examples 1.6 and 1.8 when $F(t) \equiv \sum_{j=k}^{\infty} [(\lambda t)^j / j!] e^{-\lambda t}$ ($k = 2, 3, 4, \ldots$) and investigate their results.

References

1. Ross SM (2000) Introduction to probability models. Academic Press, San Diego
2. Sahai H, Ojeda MM (2004) Analysis of variance for random models I, II. Birkhäuser, Boston
3. Zenil H (ed) (2011) Randomness through computation. World Scientific, Singapore
4. Nies A (2009) Computability and randomness. Oxford University Press, Oxford
5. Downey RG, Hirschfeldt DR (2010) Algorithmic randomness and complexity. Springer, New York
6. Osaki S (1992) Applied stochastic system modeling. Springer, Berlin
7. Nakagawa T (2011) Stochastic processes with applications to reliability theory. Springer, London
8. Barlow RE, Proschan F (1965) Mathematical theory of reliability. Wiley, New York
9. Zhao X, Nakagawa T (2012) Optimization problems of replacement first or last in reliability theory. Euro J Oper Res 223:141–149
10. Nakagawa T (2005) Maintenance theory of reliability. Springer, London
11. Nakagawa T (2008) Advanced reliability models and maintenance policies. Springer, London
12. Nakagawa T, Zhao X (2012) Optimization problems of a parallel system with a random number of units. IEEE Trans Reliab 61:543–548
13. Pinedo M (2008) Scheduling theory, algorithm and systems. Prentice Hall, Englewood Cliffs
14. Chen M, Nakagawa T (2012) Optimal scheduling of random works with reliability applications. Asia-Pacific J Oper Res, 29: 1250027 (14 pages)
15. Nakagawa T (2007) Shock and damage models in reliability theory. Springer, London
16. Zhao X, Qian C, Nakagawa T (2013) Optimal policies for cumulative damage models with maintenance last and first. Reliab Eng Syst Saf 110:50–59
17. Kunisawa K (1975) Entropy models. Nikka Giren Shuppan, Tokyo

Chapter 2
Random Age Replacement Policies

It is well known in reliability theory: (1) When the failure time is exponential, an operating unit should be replaced only at failure and (2) an optimum age replacement is nonrandom [1, p. 86]. For such questions, we introduce the shortage and excess costs, and show that both a finite replacement time and a random replacement exist even for exponential failure times.

Suppose the unit works for a job with random working times. Then, the unit is replaced before failure at a planned time T or a random time Y, whichever occurs first, which is called *replacement first*. The expected cost rate is obtained and an optimum T_F^* which minimizes it is derived analytically. It is shown naturally that when the replacement costs at time T and time Y are the same, replacement first is not better than standard replacement in which the unit is replaced only at time T. However, if the replacement cost at time Y is lower than that at time T, then replacement first would be rather than standard replacement. We give an example and discuss numerically which replacement is better [2, 3].

If the replacement cost after failure is not so high, the unit should be working as long as possible before failure. Suppose the unit is replaced before failure at time T or at time Y, whichever occurs last, which is called *replacement last*. The expected cost rate is obtained and an optimum T_L^* which minimizes it is derived analytically. It is also shown that both replacement costs at time T and Y are the same, replacement last is not better than standard replacement. Furthermore, we compare two optimum policies for replacement first and last, and show theoretically and numerically, which is better than the other according to the ratio of replacement costs [4].

Finally, it may be wise to replace an operating unit after completion of its working time even if the replacement time T comes. Suppose the unit is replaced before failure at the first completion of working times over time T, which is called *replacement overtime*. Then, an optimum replacement time T_O^* is derived and is smaller than the other optimum times of standard replacement, and replacement first and last. When both costs of standard replacement and replacement overtime are the same, standard replacement is better than replacement overtime. However, when the two costs are different, we give an example and discuss numerically which replacement is better [5].

© Springer-Verlag London 2014   19
T. Nakagawa, *Random Maintenance Policies*,
Springer Series in Reliability Engineering, DOI 10.1007/978-1-4471-6575-0_2

Throughout this chapter, suppose the unit with a failure time X $(0 < X < \infty)$ deteriorates with age and fails according to a general distribution $F(t) \equiv \Pr\{X \le t\}$ with finite mean $\mu \equiv \int_0^\infty \overline{F}(t)dt < \infty$, where $\overline{\Phi}(t) \equiv 1 - \Phi(t)$ for any distribution $\Phi(t)$. When $F(t)$ has a density function $f(t) \equiv dF(t)/dt$, i.e., $F(t) \equiv \int_0^\infty f(u)du$, the failure rate $h(t) \equiv f(t)/\overline{F}(t)$ for $F(t) < 1$ is assumed to increase strictly from $h(0) = 0$ to $h(\infty) \equiv \lim_{t \to \infty} h(t)$, except Sect. 2.1.

2.1 Random Replacement

A unit is replaced at random time Y $(0 < Y \le \infty)$ or at failure, whichever occurs first, where Y is a random variable with a general distribution $G(t)$ and is independent of the failure time X. Then, the probability that the unit is replaced at random time Y is

$$\Pr\{Y \le X\} = \int_0^\infty \overline{F}(t)\,dG(t),$$

and the probability that it is replaced at failure is

$$\Pr\{X \le Y\} = \int_0^\infty \overline{G}(t)\,dF(t).$$

Thus, the mean time to replacement is

$$\int_0^\infty t\overline{F}(t)\,dG(t) + \int_0^\infty t\overline{G}(t)\,dF(t) = \int_0^\infty \overline{G}(t)\overline{F}(t)\,dt.$$

Therefore, using the theory of a renewal reward process [6, p. 77], [7, p. 56], the expected cost rate is [1, p. 86], [8, p. 247]

$$C_A(G) = \frac{c_F \int_0^\infty \overline{G}(t)\,dF(t) + c_R \int_0^\infty G(t)\,dF(t)}{\int_0^\infty \overline{G}(t)\overline{F}(t)\,dt}, \tag{2.1}$$

where c_R = replacement cost at random time Y, and c_F = replacement cost at failure with $c_F > c_R$, which was already given in (1.24).

It has been shown [1, p. 87] that (2.1) can be written as (Problem 1 in Sect. 2.5)

$$C_A(G) = \frac{\int_0^\infty Q(t)\,dG(t)}{\int_0^\infty S(t)\,dG(t)},$$

where

$$Q(t) \equiv c_F F(t) + c_R \overline{F}(t), \quad S(t) \equiv \int_0^t \overline{F}(u)\, du.$$

Suppose that there exists a minimum value T $(0 < T \le \infty)$ of $Q(t)/S(t)$. Because

$$\frac{Q(t)}{S(t)} \ge \frac{Q(T)}{S(T)},$$

it follows that

$$\int_0^\infty Q(t)\, dG(t) \ge \frac{Q(T)}{S(T)} \int_0^\infty S(t)\, dG(t).$$

So that,

$$C_A(G) \ge \frac{Q(T)}{S(T)} = C_A(G_T),$$

where $G_T(t)$ is the degenerate distribution placing unit mass at T, i.e., $G_T(t) \equiv 1$ for $t \ge T$ and $G_T(t) \equiv 0$ for $t < T$. If $T = \infty$ then the unit is replaced only at failure. Thus, the optimum policy is nonrandom.

An optimum time T_S^* for a standard age replacement with replacement time T $(0 < T \le \infty)$ which minimizes [1, p. 87], [8, p. 74]

$$C_S(T) \equiv \frac{Q(T)}{S(T)} = \frac{c_F F(T) + c_T \overline{F}(T)}{\int_0^T \overline{F}(t)\, dt} \tag{2.2}$$

satisfies

$$h(T) \int_0^T \overline{F}(t)\, dt - F(T) = \frac{c_T}{c_F - c_T},$$

or

$$\int_0^T \overline{F}(t)[h(T) - h(t)]\, dt = \frac{c_T}{c_F - c_T}, \tag{2.3}$$

and the resulting cost rate is

$$C_S(T_S^*) = (c_F - c_T)h(T_S^*),\tag{2.4}$$

where $c_T =$ replacement cost at time T.

Example 2.1 (Random replacement for exponential failure and random times) When $F(t) = 1 - e^{-\lambda t}$ $(0 < \lambda < \infty)$ and $G(t) = 1 - e^{-\theta t}$ $(0 \le \theta < \infty)$, the expected cost rate in (2.1) becomes a function of θ and is given by

$$C_A(\theta) = c_F\lambda + c_R\theta.$$

An optimum θ^* that minimizes $C_A(\theta)$ is $\theta^* = 0$ and $C_A(0) = c_F\lambda$, i.e., we should make no random maintenance. □

Example 2.2 (Replacement for Uniform random time) Suppose that Y has a uniform distribution for the interval $[0, T]$ $(0 < T < \infty)$, i.e., $G(t) \equiv t/T$ for $t \le T$ and 1 for $t > T$. Then, the expected cost rate in (2.1) is a function of T and is given by

$$C_U(T)\frac{c_F \int_0^T F(t)\,dt + c_R \int_0^T \overline{F}(t)\,dt}{\int_0^T (T-t)\overline{F}(t)\,dt}.\tag{2.5}$$

An optimum T_U^* which minimizes $C_U(T)$ will be discussed in Sect. 2.2.4. □

2.1.1 Shortage and Excess Costs

Introduce the following two kinds of costs in Fig. 2.1 which will be more clearly defined in the scheduling problem of Chap. 7: If the unit would fail after time Y, then this causes a shortage cost $c_S(X - Y)$ because it might operate for a little more time [9, p. 83]. On the other hand, if the unit would fail before time Y, then this causes an excess cost $c_E(Y - X)$ due to its failure because it has failed at a little earlier time than Y, where $c_S(0) = c_E(0) \equiv 0$.

Under the above assumptions, the expected replacement cost is

$$
\begin{aligned}
C_1(G) &= \int_0^\infty \left[\int_0^t c_S(t-u)\,dG(u)\right]dF(t) + \int_0^\infty \left[\int_0^t c_E(t-u)\,dF(u)\right]dG(t) \\
&= \int_0^\infty \left[\int_0^t G(u)\,dc_S(u)\right]dF(t) + \int_0^\infty \left[\int_0^t F(u)\,dc_E(u)\right]dG(t) \\
&= \int_0^\infty G(t)\overline{F}(t)\,dc_S(t) + \int_0^\infty \overline{G}(t)F(t)\,dc_E(t).
\end{aligned}\tag{2.6}
$$

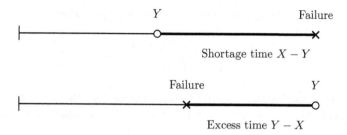

Fig. 2.1 Shortage and excess times of age replacement

Because the mean time to replacement is $\int_0^\infty \overline{G}(t)\overline{F}(t)\,dt$, the expected cost rate is

$$C_2(G) = \frac{\int_0^\infty G(t)\overline{F}(t)\,dc_S(t) + \int_0^\infty \overline{G}(t)F(t)\,dc_E(t)}{\int_0^\infty \overline{G}(t)\overline{F}(t)\,dt}. \tag{2.7}$$

When $c_S(t) = c_S t$, $c_E(t) = c_E t$, and $G(t) = 1 - e^{-\theta t}$ $(0 < \theta < \infty)$, the expected costs $C_i(G)$ $(i = 1, 2)$ are the functions of θ: From (2.6),

$$C_1(\theta) = c_S\mu + \frac{c_E}{\theta} - (c_S + c_E) \int_0^\infty e^{-\theta t}\overline{F}(t)\,dt. \tag{2.8}$$

Clearly,

$$C_1(0) \equiv \lim_{\theta \to 0} C_1(\theta) = \infty, \qquad C_1(\infty) \equiv \lim_{\theta \to \infty} C_1(\theta) = c_S\mu.$$

Differentiating $C_1(\theta)$ with respect to θ and setting it equal to zero (Problem 2 in Sect. 2.5),

$$\int_0^\infty [1 - (1 + \theta t)e^{-\theta t}]\,dF(t) = \frac{c_E}{c_S + c_E}, \tag{2.9}$$

whose left-hand side increases with θ from 0 to 1.
From (2.7),

$$C_2(\theta) = \frac{c_S\mu + c_E/\theta}{\int_0^\infty e^{-\theta t}\overline{F}(t)\,dt} - (c_S + c_E). \tag{2.10}$$

Clearly,

$$C_2(0) \equiv \lim_{\theta \to 0} C_2(\theta) = C_2(\infty) \equiv \lim_{\theta \to \infty} C_2(\theta) = \infty.$$

Differentiating $C_2(\theta)$ with respect to θ and setting it equal to zero,

$$\frac{\int_0^\infty [1-(1+\theta t)e^{-\theta t}]\,dF(t)}{\int_0^\infty t e^{-\theta t}\,dF(t)} = \frac{c_E}{c_S \mu}, \tag{2.11}$$

whose left-hand side increases strictly with θ from 0 to ∞. Therefore, there exists both finite and unique θ_i^* $(0 < \theta_i^* < \infty)$ $(i = 1, 2)$ which satisfies (2.9) and (2.11), respectively.

In particular, when $F(t) = 1 - e^{-\lambda t}$, (2.9) becomes

$$\left(\frac{\theta}{\theta+\lambda}\right)^2 = \frac{c_E}{c_S + c_E},$$

and (2.11) becomes

$$\left(\frac{\theta}{\lambda}\right)^2 = \frac{c_E}{c_S},$$

and hence, $\theta_1^* > \theta_2^*$ (Problem 3 in Sect. 2.5).

Next, when $c_S(t) = c_S t$, $c_E(t) = c_E t$, and $G(t) \equiv 0$ for $t < T$ and 1 for $t \geq T$, the expected costs $C_i(G)$ are the function of T, and from (2.6),

$$C_1(T) = c_S \int_T^\infty \overline{F}(t)\,dt + c_E \int_0^T F(t)\,dt.$$

Differentiating $C_1(T)$ with respect to T and setting it equal to zero,

$$F(T) = \frac{c_S}{c_S + c_E}. \tag{2.12}$$

The expected cost rate in (2.7) is

$$C_2(T) = \frac{c_S \int_T^\infty \overline{F}(t)\,dt + c_E \int_0^T F(t)\,dt}{\int_0^T \overline{F}(t)\,dt}.$$

Clearly, $C_2(0) = C_2(\infty) = \infty$. Differentiating $C_2(T)$ with respect to T and setting it equal to zero,

$$\frac{1}{\overline{F}(T)} \int_0^T [F(T) - F(t)]\,dt = \frac{\mu c_S}{c_E}, \tag{2.13}$$

whose left-hand side increases strictly with T from 0 to ∞. Thus, there exists a finite and unique T_2^* $(0 < T_2^* < \infty)$ which satisfies (2.13).

When $F(t) = 1 - e^{-\lambda t}$, T_1^* which satisfies (2.12) and T_2^* which satisfies (2.13) are given, respectively,

$$e^{\lambda T} - 1 = \frac{c_S}{c_E},$$

$$e^{\lambda T} - (1 + \lambda T) = \frac{c_S}{c_E},$$

whose left-hand side agrees with (8.5) of [8, p. 204] for periodic inspection with exponential failure time. In this case, $T_1^* < T_2^*$. In particular case of $c_S = c_E$, $\lambda T_1^* = \log 2 = 0.693$ and $\lambda T_2^* = 1.15$.

2.1.2 Comparison of Age and Random Replacement Policies

When $G(t) = 1 - e^{-\theta t}$ and the failure rate $h(t)$ increases strictly to $h(\infty)$, we find an optimum θ^* which minimizes the expected cost rate in (2.1) given by

$$C_A(\theta) = \frac{c_R + (c_F - c_R) \int_0^\infty e^{-\theta t}\, dF(t)}{\int_0^\infty e^{-\theta t} \overline{F}(t)\, dt}. \tag{2.14}$$

Differentiating $C_A(\theta)$ with respect to θ and setting it equal to zero,

$$Q_1(\theta) \int_0^\infty e^{-\theta t} \overline{F}(t)\, dt - \int_0^\infty e^{-\theta t} dF(t) = \frac{c_R}{c_F - c_R}, \tag{2.15}$$

where $Q_1(\theta) \equiv \lim_{T \to \infty} Q_1(T; \theta)$ and for $0 < T \le \infty$,

$$Q_1(T; \theta) \equiv \frac{\int_0^T t e^{-\theta t}\, dF(t)}{\int_0^T t e^{-\theta t} \overline{F}(t)\, dt} \le h(T).$$

First, note from (5) of Appendix A.1 that $h(0) \le Q_1(T; \theta) \le h(T)$ and $Q_1(T; \theta)$ increases strictly with T from $h(0)$ to

$$Q_1(\theta) = \frac{\int_0^\infty t e^{-\theta t}\, dF(t)}{\int_0^\infty t e^{-\theta t} \overline{F}(t)\, dt}.$$

Furthermore, $Q_1(T; \theta)$ decreases strictly with θ from

$$Q_1(T; 0) \equiv \lim_{\theta \to 0} Q_1(T; \theta) = \frac{\int_0^T t \, dF(t)}{\int_0^T t \overline{F}(t) \, dt}$$

to $h(0) = 0$.

From the above results, the left-hand side of (2.15) decreases strictly with θ from

$$Q_1(0)\mu - 1 = \frac{\mu \int_0^\infty t \, dF(t)}{\int_0^\infty t \overline{F}(t) \, dt} - 1 = \frac{2\mu \int_0^\infty t \, dF(t)}{\int_0^\infty t^2 dF(t)} - 1 = \frac{2\mu^2}{\mu^2 + \sigma^2} - 1$$

to 0, where $\sigma^2 \equiv \int_0^\infty (t - \mu)^2 dF(t)$. Therefore, if $Q_1(0) > c_F/[\mu(c_F - c_R)]$, i.e., $2\mu^2/(\mu^2 + \sigma^2) > c_F/(c_F - c_R)$, then there exists an optimum θ^* ($0 < \theta^* < \infty$) which satisfies (2.15), and the resulting cost rate is

$$C_A(\theta^*) = (c_F - c_R)Q_1(\theta^*). \tag{2.16}$$

Note that $Q_1(\theta)$ plays the same role as the failure rate $h(t)$ in standard replacement.

We have already known that if $c_T \leq c_R$, then standard replacement with replacement time T is better than random one in Sect. 2.1. When $c_T > c_R$ and $Q_1(0) > c_F/[\mu(c_F - c_R)]$, we compare the expected cost rates $C_S(T)$ in (2.2) and $C_A(\theta)$ in (2.14). We derive a replacement cost $\widehat{c}_R$ in which two optimum cost rates $C_S(T_S^*)$ in (2.4) and $C_A(\theta^*)$ in (2.16) are the same. First, we compute T_S^* ($0 < T_S^* < \infty$) which satisfies (2.3) for c_T and c_F, and $C_S(T_S^*)$ in (2.4). Next, we compute $\widehat{c}_R$ which satisfies

$$Q_1(\theta) \int_0^\infty e^{-\theta t} \overline{F}(t) \, dt + \int_0^\infty (1 - e^{-\theta t}) dF(t) = \frac{c_F}{c_F - \widehat{c}_R},$$

$$(c_F - c_T)h(T_S^*) = (c_F - \widehat{c}_R)Q_1(\theta),$$

i.e., we firstly obtain $\widehat{\theta}$ which satisfies

$$\frac{1}{Q_1(\widehat{\theta})} \int_0^\infty (1 - e^{-\widehat{\theta}t}) dF(t) + \int_0^\infty e^{-\widehat{\theta}t} \overline{F}(t) \, dt = \frac{1}{h(T_S^*)} \frac{c_F}{c_F - c_T}, \tag{2.17}$$

and using $\widehat{\theta}$, we compute $\widehat{c}_R$ which satisfies

$$\frac{c_F - \widehat{c}_R}{c_F - c_T} = \frac{h(T_S^*)}{Q_1(\widehat{\theta})}. \tag{2.18}$$

When $T_S^* = \infty$, $\widehat{\theta} = 0$ and $C_S(\infty) = C_A(0) = c_F/\mu$, and hence, $\widehat{c}_R = c_T$.

Example 2.3 (*Random replacement for gamma failure time*) Suppose that the failure time has a gamma distribution with order k, i.e., $F(t) = \sum_{j=k}^{\infty} [(\lambda t)^j / j!] e^{-\lambda t}$ ($k = 2, 3, \ldots$), $f(t) = [\lambda(\lambda t)^{k-1} / (k-1)!] e^{-\lambda t}$, $\mu = k/\lambda$ and $h(t) = [\lambda(\lambda t)^{k-1} / (k-1)!] / \sum_{j=0}^{k-1} [(\lambda t)^j / j!]$, which increases strictly with t from 0 to λ. Then, if $k > c_F / (c_F - c_T)$, then an optimum T_S^* ($0 < T_S^* < \infty$) satisfies uniquely

$$\frac{\lambda(\lambda T)^{k-1} / (k-1)!}{\sum_{j=0}^{k-1} [(\lambda T)^j / j!]} \sum_{j=0}^{k-1} \int_0^T \frac{(\lambda t)^j}{j!} e^{-\lambda t} \, dt + \sum_{j=0}^{k-1} \frac{(\lambda T)^j}{j!} e^{-\lambda T} = \frac{c_F}{c_F - c_T},$$

and the resulting cost rate is

$$C_S(T_S^*) = (c_F - c_T) \frac{\lambda(\lambda T_S^*)^{k-1} / (k-1)!}{\sum_{j=0}^{k-1} [(\lambda T_S^*)^j / j!]}.$$

On the other hand, when $F(t)$ is a gamma distribution,

$$\int_0^{\infty} t e^{-\theta t} \, dF(t) = \frac{k}{\theta + \lambda} \left(\frac{\lambda}{\theta + \lambda} \right)^k,$$

$$\int_0^{\infty} t e^{-\theta t} \overline{F}(t) \, dt = \frac{1}{\lambda^2} \sum_{j=1}^{k} j \left(\frac{\lambda}{\theta + \lambda} \right)^{j+1},$$

$$Q_1(\theta) = \frac{k\lambda[\lambda/(\theta + \lambda)]^{k-1}}{\sum_{j=1}^{k} j[\lambda/(\theta + \lambda)]^{j-1}},$$

which decreases strictly with θ from $2\lambda/(k+1)$ to 0. If $2k/(k+1) > c_F/(c_F - c_R)$, then from (2.15), an optimum θ^* ($0 < \theta^* < \infty$) satisfies uniquely

$$\frac{kA^{k-1}}{\sum_{j=1}^{k} jA^{j-1}} \sum_{j=1}^{k} A^j - A^k = \frac{c_R}{c_F - c_R},$$

where $A \equiv \lambda/(\theta + \lambda)$, and the resulting cost rate is

$$C_A(\theta^*) = (c_F - c_R) \frac{k\lambda[\lambda/(\theta^* + \lambda)]^{k-1}}{\sum_{j=1}^{k} j[\lambda/(\theta^* + \lambda)]^{j-1}}.$$

Next, we compute a replacement cost $\widehat{c}_R$ for random replacement in which both expected cost rates of age and random replacement are the same. Using T_S^*, we compute $\widehat{\theta}$ which satisfies (2.17), and using both T_S^* and $\widehat{\theta}$, we compute $\widehat{c}_R/c_F$ from (2.18). If $2k/(k+1) \le c_F/(c_F - c_R)$, then $T_S^* = 1/\theta^* = \infty$, and hence, $\widehat{c}_R = c_T$.

Table 2.1 Optimum T_S^*, $1/\theta^*$, $1/\widehat{\theta}$, $\widehat{c}_R/c_F$, and $\widehat{c}_R/c_T$ when $F(t) = \sum_{j=k}^{\infty}(t^j/j!)e^{-t}$

c_T/c_F	$k=2$					$k=3$				
or c_R/c_F	T_S^*	$1/\theta^*$	$1/\widehat{\theta}$	$\widehat{c}_R/c_F$	$\widehat{c}_R/c_T$	T_S^*	$1/\theta^*$	$1/\widehat{\theta}$	$\widehat{c}_R/c_F$	$\widehat{c}_R/c_T$
0.01	0.157	0.051	0.030	0.006	0.600	0.357	0.195	0.137	0.005	0.500
0.02	0.233	0.062	0.044	0.012	0.600	0.468	0.299	0.204	0.010	0.500
0.05	0.412	0.335	0.207	0.031	0.620	0.697	0.540	0.348	0.026	0.520
0.1	0.680	0.793	0.430	0.064	0.640	0.984	0.955	0.557	0.053	0.530
0.2	1.306	4.017	1.104	0.129	0.645	1.512	2.462	1.042	0.109	0.545

In other words, if c_F approaches to c_T and c_R then both T_S^* and $1/\theta^*$ become large and $\widehat{c}_R/c_T$ approaches to 1.

Table 2.1 presents T_S^*, $1/\theta^*$, $1/\widehat{\theta}$, $\widehat{c}_R/c_F$ and $\widehat{c}_R/c_T$ for c_T/c_F or c_R/c_F when $F(t) = \sum_{j=k}^{\infty}(t^j/j!)e^{-t}$ ($k = 2, 3$). From the comparison results among T_S^*, $1/\theta^*$ and $1/\widehat{\theta}$, $T_S^* > 1/\theta^*$ for small c_T/c_F or c_R/c_F, however, $T_S^* < 1/\theta^*$ for large ones. Note that $1/\widehat{\theta} < 1/\theta^*$ and has the same variation trend with $1/\theta^*$. It is also shown that $\widehat{c}_R/c_F$ decreases with k and increases with c_T/c_F or c_R/c_F. From the numerical values of $\widehat{c}_R/c_T$, we can find that if how much $\widehat{c}_R$ is less than c_T, the expected costs of standard and random replacements are the same. Taking $k = 2$ for example, when $\widehat{c}_R$ is a little higher than 60 % of c_T, we should adopt random replacement. $\square$

2.2 Random Replacement Policies

Using age and random replacement policies introduced in Sect. 2.1, we define three new policies of replacement first, replacement last, and replacement overtime, and discuss their optimum policies which minimize the expected cost rates. Furthermore, we compare the three policies with standard replacement.

2.2.1 Replacement First

When the failure rate $h(t)$ increases strictly to $h(\infty)$, we consider an age replacement policy in which the unit is replaced at time T, Y or at failure, whichever occurs first, i.e., at time min $\{T, X, Y\}$. This is called *replacement first*. It is assumed that T ($0 < T \le \infty$) is constant, and X and Y are independent random variables with the respective distributions $F(t)$ and $G(t)$. Then, the probability that the unit is replaced at time T is

$$\Pr\{Y > T, X > T\} = \overline{G}(T)\overline{F}(T), \tag{2.19}$$

the probability that it is replaced at random time Y is

$$\Pr\{Y \leq T, Y \leq X\} = \int_0^T \overline{F}(t)\,dG(t), \qquad (2.20)$$

and the probability that it is replaced at failure is

$$\Pr\{X \leq T, X \leq Y\} = \int_0^T \overline{G}(t)\,dF(t), \qquad (2.21)$$

where $(2.19) + (2.20) + (2.21) = 1$. Thus, the mean time to replacement is

$$T\overline{G}(T)\overline{F}(T) + \int_0^T t\overline{F}(t)\,dG(t) + \int_0^T t\overline{G}(t)\,dF(t) = \int_0^T \overline{G}(t)\overline{F}(t)\,dt. \qquad (2.22)$$

Therefore, the expected cost rate is

$$C_F(T) = \frac{c_T + (c_F - c_T)\int_0^T \overline{G}(t)\,dF(t) + (c_R - c_T)\int_0^T \overline{F}(t)\,dG(t)}{\int_0^T \overline{G}(t)\overline{F}(t)\,dt}, \qquad (2.23)$$

where c_F = replacement cost at failure, and c_T and c_R are the respective replacement costs at time T and random time Y with $c_F > c_T$ and $c_F > c_R$. Clearly,

$$C_F(0) \equiv \lim_{T \to 0} C_F(T) = \infty,$$

$$C_F(\infty) \equiv \lim_{T \to \infty} C_F(T) = \frac{c_F - (c_F - c_R)\int_0^\infty \overline{F}(t)\,dG(t)}{\int_0^\infty \overline{G}(t)\overline{F}(t)\,dt},$$

which agrees with (2.1). First, suppose that $Y = \infty$, i.e., $\overline{G}(t) \equiv 1$ for $t \geq 0$. Then, the unit cannot be replaced at random time Y, and the expected cost rate is

$$C_S(T) = \frac{c_T + (c_F - c_T)F(T)}{\int_0^T \overline{F}(t)\,dt}, \qquad (2.24)$$

which agrees with (2.2). Therefore, if $h(\infty) > c_F/[\mu(c_F - c_T)]$, then there exists an optimum T_S^* $(0 < T_S^* < \infty)$ which satisfies (2.3), and the resulting cost rate is given in (2.4)

Next, when $c_T = c_R$, the expected cost rate is, from (2.23),

$$C_F(T) = \frac{c_T + (c_F - c_T) \int_0^T \overline{G}(t) dF(t)}{\int_0^T \overline{G}(t) \overline{F}(t) dt}. \tag{2.25}$$

We find an optimum T_F^* which minimizes $C_F(T)$. Differentiating $C_F(T)$ with respect to T and setting it equal to zero,

$$h(T) \int_0^T \overline{G}(t) \overline{F}(t) dt - \int_0^T \overline{G}(t) dF(t) = \frac{c_T}{c_F - c_T}, \tag{2.26}$$

whose left-hand side increases strictly from 0 because $h(t)$ increases strictly to $h(\infty)$. Thus, if

$$h(\infty) \int_0^\infty \overline{G}(t) \overline{F}(t) dt - \int_0^\infty \overline{G}(t) dF(t) > \frac{c_T}{c_F - c_T},$$

then there exists a finite and unique T_F^* $(0 < T_F^* < \infty)$ which satisfies (2.26), and the resulting cost rate is

$$C_F(T_F^*) = (c_F - c_T) h(T_F^*). \tag{2.27}$$

In addition, (2.26) is written as

$$\int_0^T \overline{G}(t) \overline{F}(t) [h(T) - h(t)] dt = \frac{c_T}{c_F - c_T}. \tag{2.28}$$

In particular, when $G(t) = 1 - e^{-\theta t}$ $(0 < \theta < \infty)$, (2.28) becomes

$$\int_0^T e^{-\theta t} \overline{F}(t) [h(T) - h(t)] dt = \frac{c_T}{c_F - c_T}. \tag{2.29}$$

Thus, an optimum T_F^* increases with θ from T_S^* given in (2.3), i.e., decreases with mean random time $1/\theta$, and $T_F^* > T_S^*$ (Problem 4 in Sect. 2.5). This means that if $1/\theta$ increases, then the unit should be replaced mainly at time T, and hence, T_F^* decreases. In addition, if

$$h(\infty) \frac{1 - F^*(\theta)}{\theta} - F^*(\theta) > \frac{c_T}{c_F - c_T},$$

then there exists a finite and unique T_F^* ($0 < T_F^* < \infty$) which satisfies (2.29), where $F^*(s)$ is the LS (Laplace-Stieltjes) transform of $F(t)$, i.e., $F^*(s) \equiv \int_0^\infty e^{-st} dF(t)$ for $Re(s) > 0$.

Furthermore, from (2.3) and (2.28),

$$h(T) \int_0^T \overline{F}(t)\, dt - F(T) \geq \int_0^T \overline{G}(t)\overline{F}(t)[h(T) - h(t)]\, dt$$

follows that $T_F^* \geq T_S^*$. So that, comparing (2.4) with (2.27), $C_S(T_S^*) \leq C_F(T_F^*)$. Thus, standard replacement with only time T is better than replacement first, as already shown in Sect. 2.1. However, if the replacement cost c_R at time Y would be lower than c_T, then replacement first might be rather than standard replacement.

It is assumed from the above discussions that $c_T > c_R$. Differentiating $C_F(T)$ in (2.23) with respect to T and setting it equal to zero,

$$(c_F - c_T)\left[h(T) \int_0^T \overline{G}(t)\overline{F}(t)\, dt - \int_0^T \overline{G}(t) dF(t) \right]$$

$$- (c_T - c_R)\left[r(T) \int_0^T \overline{G}(t)\overline{F}(t)\, dt - \int_0^T \overline{F}(t) dG(t) \right] = c_T, \qquad (2.30)$$

where $r(t) \equiv g(t)/\overline{G}(t)$ and $g(t)$ is a density function of $G(t)$. If $r(t)$ decreases with t, then the left-hand side of (2.30) increases strictly with T from 0 to

$$(c_F - c_T) \int_0^\infty \overline{G}(t)\overline{F}(t)[h(\infty) - h(t)]\, dt$$

$$- (c_T - c_R) \int_0^T \overline{G}(t)\overline{F}(t)[r(\infty) - r(t)]\, dt. \qquad (2.31)$$

Thus, if (2.31) is greater than c_T, then there exists a finite and unique T_F^* ($0 < T_F^* < \infty$) which satisfies (2.30), and the resulting cost rate is

$$C_F(T_F^*) = (c_F - c_T)h(T_F^*) - (c_T - c_R)r(T_F^*). \qquad (2.32)$$

In particular, when $G(t) = 1 - e^{-\theta t}$, i.e., $r(t) = \theta$, (2.30) agrees with (2.29).

2.2.2 Replacement Last

Suppose that the unit is replaced at time T $(0 \leq T < \infty)$ or at time Y, whichever occurs last, or at failure, i.e., it is replaced at time max $\{T, Y\}$ or at failure, whichever occurs first. This is called *replacement last*. Then, the probability that the unit is replaced at time T is

$$\Pr\{Y \leq T < X\} = G(T)\overline{F}(T), \tag{2.33}$$

the probability that it is replaced at time Y is

$$\Pr\{T < Y < X\} = \int_T^\infty \overline{F}(t)\,\mathrm{d}G(t), \tag{2.34}$$

and the probability that it is replaced at failure is

$$\Pr\{X \leq T \text{ or } T < X \leq Y\} = F(T) + \int_T^\infty \overline{G}(t)\,\mathrm{d}F(t), \tag{2.35}$$

where $(2.33) + (2.34) + (2.35) = 1$. Thus, the mean time to replacement is

$$TG(T)\overline{F}(T) + \int_T^\infty t\overline{F}(t)\,\mathrm{d}G(t) + \int_0^T t\,\mathrm{d}F(t) + \int_T^\infty t\overline{G}(t)\,\mathrm{d}F(t)$$

$$= \int_0^T \overline{F}(t)\,\mathrm{d}t + \int_T^\infty \overline{G}(t)\overline{F}(t)\,\mathrm{d}t. \tag{2.36}$$

Therefore, the expected cost rate is

$$C_L(T) = \frac{c_F - (c_F - c_T)G(T)\overline{F}(T) - (c_F - c_R)\int_T^\infty \overline{F}(t)\,\mathrm{d}G(t)}{\int_0^T \overline{F}(t)\,\mathrm{d}t + \int_T^\infty \overline{G}(t)\overline{F}(t)\,\mathrm{d}t}, \tag{2.37}$$

where c_F, c_T, and c_R are given in (2.23). Clearly,

$$C_L(0) \equiv \lim_{T \to 0} C_L(T) = \frac{c_F - (c_F - c_R)\int_0^\infty \overline{F}(t)\mathrm{d}G(t)}{\int_0^\infty \overline{G}(t)\overline{F}(t)\,\mathrm{d}t} = C_F(\infty),$$

$$C_L(\infty) \equiv \lim_{T \to \infty} C_L(T) = \frac{c_F}{\mu} = C_S(\infty),$$

which is the expected cost rate when the unit is replaced only at failure. In particular, when $Y = 0$, i.e., $G(t) \equiv 1$ for $t \geq 0$, $C_L(T)$ agrees with $C_S(T)$ in (2.2).

Next, when $c_T = c_R$, the expected cost rate is, from (2.37),

$$C_L(T) = \frac{c_F - (c_F - c_T) \int_T^\infty G(t) dF(t)}{\int_0^T \overline{F}(t) dt + \int_T^\infty \overline{G}(t)\overline{F}(t) dt}. \tag{2.38}$$

We find an optimum T_L^* which minimizes $C_L(T)$. Differentiating $C_L(T)$ with respect to T and setting it equal to zero,

$$h(T)\left[\int_0^T \overline{F}(t)dt + \int_T^\infty \overline{G}(t)\overline{F}(t)dt\right] - \left[1 - \int_T^\infty G(t)dF(t)\right] = \frac{c_T}{c_F - c_T}, \tag{2.39}$$

whose left-hand side increases strictly to $\mu h(\infty) - 1$. Thus, if $h(\infty) > c_F/[\mu(c_F - c_T)]$, then there exists a finite and unique T_L^* ($0 \leq T_L^* < \infty$) which satisfies (2.39), and the resulting cost rate is

$$C_L(T_L^*) = (c_F - c_T)h(T_L^*). \tag{2.40}$$

Note that if a finite T_F^* in (2.26) exists, then both finite T_S^* in (2.3) and T_L^* in (2.39) exist. In particular, when $G(t) = 1 - e^{-\theta t}$, (2.39) becomes

$$\int_0^T \overline{F}(t)[h(T) - h(t)] dt - \int_T^\infty e^{-\theta t}\overline{F}(t)[h(t) - h(T)] dt = \frac{c_T}{c_F - c_T}.$$

Thus, an optimum T_L^* decreases with θ to T_S^* given in (2.3), i.e., increases with a mean random time $1/\theta$, and $T_L^* > T_S^*$ (Problem 5 in Sect. 2.5).

Furthermore, from (2.3) and (2.39),

$$h(T)\int_0^T \overline{F}(t)dt - F(T) \geq h(T)\left[\int_0^T \overline{F}(t)dt + \int_T^\infty \overline{G}(t)\overline{F}(t)dt\right] + \int_T^\infty G(t)dF(t) - 1,$$

follows that $T_L^* \geq T_S^*$. So that, comparing (2.4) with (2.40), $C_S(T_S^*) \leq C_L(T_L^*)$. Thus, standard replacement is better than replacement last, as shown similarly in Sect. 2.2.1. However, if the replacement cost c_R at time Y would be lower than c_T, then replacement last might be rather than standard replacement.

It is assumed from the above discussions that $c_T > c_R$. Differentiating $C_L(T)$ in (2.37) with respect to T and setting it equal to zero,

$$
(c_F - c_T) \left\{ h(T) \left[\int_0^T \overline{F}(t)\,dt + \int_T^\infty \overline{G}(t)\overline{F}(t)\,dt \right] - F(T) \right.
$$

$$
\left. - \int_T^\infty \overline{G}(t)\,dF(t) \right\} + (c_T - c_R) \left\{ \tilde{r}(T) \left[\int_0^T \overline{F}(t)\,dt \right. \right.
$$

$$
\left. \left. + \int_T^\infty \overline{G}(t)\overline{F}(t)\,dt \right] + \int_T^\infty \overline{F}(t)\,dG(t) \right\} = c_T, \tag{2.41}
$$

where $\tilde{r}(t) \equiv g(t)/\overline{G}(t)$. Denoting left-hand side of (2.41) by $L_1(T)$,

$$
L_1'(T) = [(c_F - c_T)h'(T) + (c_T - c_R)\tilde{r}'(T)] \left[\int_0^T \overline{F}(t)\,dt + \int_T^\infty \overline{G}(t)\overline{F}(t)\,dt \right],
$$

$$
L_1(\infty) = (c_F - c_T)[h(\infty)\mu - 1] + (c_T - c_R)\tilde{r}(\infty)\mu.
$$

Therefore, if $(c_F - c_T)h(t) + (c_T - c_R)\tilde{r}(t)$ increases strictly with t and $L_1(\infty) > c_T$, i.e., $(c_F - c_T)h(\infty) + (c_T - c_R)\tilde{r}(\infty) > c_F/\mu$ then there exists a finite and unique T_L^* ($0 \le T_L^* < \infty$) which satisfies (2.41). Note that when $G(t) = 1 - e^{-\theta t}$, $\tilde{r}(t)$ decreases from ∞ to 0. In this case, if $h(\infty) > c_F/[\mu(c_F - c_T)]$ then a finite T_L^* to satisfy (2.41) exists.

2.2.3 Replacement Overtime

It might be wise to replace practically the unit at the completion of its working time even if T comes because it continues to work for some job. Suppose in the age replacement that the unit is replaced before failure at the first completion of random times Y_j ($j = 1, 2, \ldots$) over time T ($0 \le T < \infty$), where Y_j is independent and has an identical distribution $G(t) \equiv \Pr\{Y_j \le j\}$ with finite mean $1/\theta$ in Fig. 2.2, where $G^{(j)}(t)$ ($j = 1, 2, \ldots$) is the j-fold Stieltjes convolution of $G(t)$ with itself, and $G^{(0)}(t) \equiv 1$ for $t \ge 0$. This is called *replacement overtime* [5].

The probability that the unit is replaced before failure at the first completion of working times over time T is

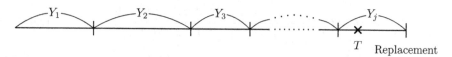

Fig. 2.2 Replacement over time T

$$\sum_{j=0}^{\infty} \int_0^T \left[\int_{T-t}^{\infty} \overline{F}(t+u) dG(u) \right] dG^{(j)}(t), \qquad (2.42)$$

the probability that it is replaced at failure before time T is

$$\sum_{j=0}^{\infty} \int_0^T [G^{(j)}(t) - G^{(j+1)}(t)] dF(t) = F(T), \qquad (2.43)$$

and the probability that it is replaced at failure after time T is

$$\sum_{j=0}^{\infty} \int_0^T \left\{ \int_{T-t}^{\infty} [F(t+u) - F(T)] dG(u) \right\} dG^{(j)}(t), \qquad (2.44)$$

where $(2.42) + (2.43) + (2.44) = 1$ (Problem 6 in Sect. 2.5). Thus, the mean time to replacement is

$$\sum_{j=0}^{\infty} \int_0^T \left[\int_{T-t}^{\infty} (t+u) \overline{F}(t+u) dG(u) \right] dG^{(j)}(t) + \int_0^T t \, dF(t)$$

$$+ \sum_{j=0}^{\infty} \int_0^T \left\{ \int_{T-t}^{\infty} \left[\int_T^{t+u} v \, dF(v) \right] dG(u) \right\} dG^{(j)}(t),$$

$$= \int_0^T \overline{F}(t) \, dt + \sum_{j=0}^{\infty} \int_0^T \left[\int_T^{\infty} \overline{G}(u-t) \overline{F}(u) du \right] dG^{(j)}(t). \qquad (2.45)$$

Therefore, the expected cost rate is

$$C_O(T) = \frac{c_F - (c_F - c_R) \sum_{j=0}^{\infty} \int_0^T [\int_{T-t}^{\infty} \overline{F}(t+u) dG(u)] dG^{(j)}(t)}{\int_0^T \overline{F}(t) \, dt + \sum_{j=0}^{\infty} \int_0^T [\int_T^{\infty} \overline{G}(u-t) \overline{F}(u) du] dG^{(j)}(t)}, \qquad (2.46)$$

where $c_R =$ replacement cost over time T and c_F is given in (2.23).

In particular, when $G(t) = 1 - e^{-\theta t}$,

$$C_O(T) = \frac{c_F - (c_F - c_R) \int_T^\infty \overline{F}(t)\theta e^{-\theta(t-T)}\, dt}{\int_0^T \overline{F}(t)\, dt + \int_T^\infty \overline{F}(t)e^{-\theta(t-T)}\, dt}. \qquad (2.47)$$

Clearly,

$$C_O(0) \equiv \lim_{T \to 0} C_O(T) = \frac{c_F - (c_F - c_R) \int_0^\infty \overline{F}(t)\theta e^{-\theta t}\, dt}{\int_0^\infty \overline{F}(t)e^{-\theta t}\, dt} = C_F(\infty) = C_L(0),$$

$$C_O(\infty) \equiv \lim_{T \to \infty} C_O(T) = \frac{c_F}{\mu} = C_S(\infty) = C_L(\infty).$$

We find an optimum T_O^* which minimizes $C_O(T)$ in (2.47). Differentiating $C_O(T)$ with respect to T and setting it equal to zero,

$$\widetilde{Q}_0(T; \theta) \int_0^T \overline{F}(t)\, dt - F(T) = \frac{c_R}{c_F - c_R}, \qquad (2.48)$$

where

$$\widetilde{Q}_0(T; \theta) \equiv \frac{\int_T^\infty e^{-\theta t} dF(t)}{\int_T^\infty e^{-\theta t}\overline{F}(t)\, dt} \geq h(T).$$

First, note from (7) of Appendix A.1 that when the failure rate $h(t)$ increases strictly, $\widetilde{Q}_0(T; \theta)$ is greater than $h(T)$ and increases strictly to $h(\infty)$, and increases with θ to $h(T)$ for $0 \leq T < \infty$.

From the above results, if $h(\infty) > c_F/[\mu(c_F - c_R)]$, then there exists a finite and unique T_O^* $(0 < T_O^* < \infty)$ which satisfies (2.48), and the resulting cost rate is (Problem 7 in Sect. 2.5)

$$C_O(T_O^*) = (c_F - c_R)\widetilde{Q}_0(T_O^*; \theta) = \frac{c_R + (c_F - c_R)F(T_O^*)}{\int_0^{T_O^*} \overline{F}(t)\, dt}. \qquad (2.49)$$

Because $\widetilde{Q}_0(T; \theta)$ decreases strictly with θ from $\overline{F}(T)/\int_T^\infty \overline{F}(t) dt$ to $h(T)$, T_O^* increases with θ to T_S^* given in (2.3) when $c_T = c_R$ and $T_O^* > T_S^*$. Standard replacement is better than replacement overtime. This is proved easily from (2.2) and (2.49) because T_S^* minimizes the right-hand side of (2.49).

2.2.4 Replacement of Uniform Random Time

We find an optimum T_U^* which minimizes $C_U(T)$ in (2.5). Clearly,

$$C_U(0) \equiv \lim_{T \to 0} C_U(T) = \lim_{T \to 0} C_S(T) = \infty,$$

$$C_U(\infty) \equiv \lim_{T \to \infty} C_U(T) = \lim_{T \to \infty} C_S(T) = \frac{c_F}{\mu}.$$

Differentiating $C_U(T)$ with respect to T and setting it equal to zero,

$$Q_1(T) \int_0^T \overline{F}(t)\,dt - F(T) = \frac{c_R}{c_F - c_R}, \qquad (2.50)$$

where

$$Q_1(T) \equiv Q_1(T;0) = \frac{\int_0^T t\,dF(t)}{\int_0^T t\overline{F}(t)\,dt} \le h(T).$$

First, note that when the failure rate $h(t)$ increases strictly, from (**5**) of Appendix A.1, $Q_1(T)$ is less than $h(T)$ and increases strictly to

$$Q_1(\infty) = \frac{2\int_0^\infty t\,dF(t)}{\int_0^\infty t^2\,dF(t)} = \frac{2\mu}{\sigma^2 + \mu^2},$$

where $\sigma^2 \equiv V\{X\} = \int_0^\infty (t - \mu)^2\,dF(t)$. Differentiating the left-hand side of (2.50) with respect to T,

$$\frac{\overline{F}(T)}{[\int_0^T t\overline{F}(t)\,dt]^2} \int_0^T (T - t)\overline{F}(t)\,dt \int_0^T t\overline{F}(t)[h(T) - h(t)]\,dt > 0.$$

Thus, the left-hand side of (2.50) increases strictly from 0 to $\mu Q_1(\infty) - 1$. Therefore, if $Q_1(\infty) > c_F/[\mu(c_F - c_R)]$ then there exists a finite and unique T_U^* $(0 < T_U^* < \infty)$ which satisfies (2.50), and the resulting cost rate is

$$C_U(T_U^*) = (c_F - c_R)Q_1(T_U^*) = \frac{c_R + (c_F - c_R)F(T_U^*)}{\int_0^{T_U^*} \overline{F}(t)\,dt}. \qquad (2.51)$$

Because $Q_1(T) < h(T)$ for $0 < T < \infty$, $T_U^* > T_S^*$ when $c_T = c_R$, and hence, standard replacement is better than this replacement.

2.3 Comparisons of Replacement Times

We compare replacement first, replacement last, and replacement overtime analytically and numerically.

2.3.1 Comparison of Replacement First and Last

Compare the expected cost rates $C_F(T)$ in (2.25) and $C_L(T)$ in (2.38). It is assumed that the failure rate $h(t)$ increases strictly from 0 to ∞, i.e., $h(\infty) = \infty$. Then, there exists both finite and unique T_F^* $(0 < T_F^* < \infty)$ and T_L^* $(0 < T_L^* < \infty)$ which satisfies (2.26) and (2.39), respectively. First, compare the left-hand sides of (2.26) and (2.39): Denote

$$L_2(T) \equiv h(T) \left[\int_0^T \overline{F}(t)\, dt + \int_T^\infty \overline{G}(t)\overline{F}(t)\, dt \right] - \left[1 - \int_T^\infty G(t)\, dF(t) \right]$$

$$- \left[h(T) \int_0^T \overline{G}(t)\overline{F}(t)\, dt - \int_0^T \overline{G}(t)\, dF(t) \right]$$

$$= \int_0^T G(t)\overline{F}(t)[h(T) - h(t)]\, dt - \int_T^\infty \overline{G}(t)\overline{F}(t)[h(t) - h(T)]\, dt.$$

Clearly,

$$L_2(0) \equiv \lim_{T \to 0} L_2(T) = - \int_0^\infty \overline{G}(t)\, dF(t) < 0, \qquad L_2(\infty) \equiv \lim_{T \to \infty} L_2(T) = \infty,$$

$$L_2'(T) = h'(T) \left[\int_0^T G(t)\overline{F}(t)\, dt + \int_T^\infty \overline{G}(t)\overline{F}(t)\, dt \right] > 0.$$

Thus, there exists a finite and unique T_A $(0 < T_A < \infty)$ which satisfies $L_2(T) = 0$. We set that

$$L(T_A) \equiv h(T_A) \int_0^{T_A} \overline{G}(t)\overline{F}(t)\, dt - \int_0^{T_A} \overline{G}(t)\, dF(t). \qquad (2.52)$$

Then, it is shown from (2.26) to (2.39) that if $L(T_A) \geq c_T/(c_F - c_T)$, i.e., $c_F/c_T \geq 1 + 1/L(T_A)$, then $T_F^* \leq T_L^*$, and hence, from (2.27) and (2.40), replacement first is better than replacement last. Conversely, if $L(T_A) < c_T/(c_F - c_T)$, then $T_F^* > T_L^*$ and replacement last is better than replacement first. This means that if the ratio of c_T/c_F is greater than $L(T_A)/[1 + L(T_A)]$, i.e., the replacement cost c_F is nearly to cost c_T, replacement last is better than replacement first.

Table 2.2 Optimum T_F^*, T_L^*, T_A and $L(T_A)$ when $F(t) = 1 - e^{-t^2}$

$\frac{c_T}{c_F - c_T}$	$1/\theta = 0.1$		$1/\theta = 0.3$		$1/\theta = 0.5$		$1/\theta = 0.7$		$1/\theta = 1.0$		$1/\theta = \infty$
	T_F^*	T_L^*	T_F^*	T_L^*	T_F^*	T_L^*	T_F^*	T_L^*	T_F^*	T_L^*	T_S^*
0.01	0.12	0.13	0.11	0.23	0.10	0.31	0.10	0.36	0.10	0.41	0.10
0.03	0.24	0.18	0.19	0.26	0.18	0.33	0.18	0.38	0.18	0.42	0.17
0.05	0.35	0.23	0.26	0.29	0.24	0.35	0.24	0.40	0.23	0.44	0.22
0.07	0.45	0.27	0.31	0.32	0.29	0.37	0.28	0.42	0.28	0.46	0.26
0.10	0.61	0.32	0.39	0.36	0.36	0.41	0.35	0.44	0.34	0.48	0.32
0.30	1.62	0.56	0.80	0.57	0.69	0.59	0.65	0.61	0.62	0.63	0.56
0.50	2.64	0.74	1.19	0.74	0.97	0.75	0.89	0.76	0.84	0.77	0.74
0.70	3.66	0.89	1.57	0.89	1.24	0.89	1.12	0.90	1.04	0.91	0.89
1.00	5.19	1.09	2.14	1.09	1.64	1.09	1.45	1.10	1.33	1.10	1.09
T_A	0.13		0.33		0.46		0.55		0.65		
$L(T_A)$	0.011		0.077		0.155		0.228		0.325		

Example 2.4 (Replacement for Weibull failure time) Suppose that the failure time X has a Weibull distribution $F(t) = 1 - \exp(-t^2)$ and a random working time Y has an exponential distribution $G(t) = 1 - e^{-\theta t}$. Table 2.2 presents optimum T_F^*, T_L^*, $L(T_A)$, and T_S^* which satisfy (2.26), (2.39), (2.52) and (2.3), respectively, for $1/\theta$ and $c_T/(c_F - c_T)$. When $1/\theta = \infty$, T_F^* becomes equal to T_S^*.

Table 2.2 indicates that both T_F^* and T_L^* increase with $c_T/(c_F - c_T)$, i.e., decrease with c_F/c_T. When c_F/c_T increases, the replacement time should be smaller to prevent a high replacement cost. In other words, the unit can work longer as c_F/c_T becomes smaller. Furthermore, replacement last is much better than replacement first when c_F/c_T becomes smaller, especially for small $1/\theta$. For example, when $c_T/(c_F - c_T) = 0.7$, i.e., $c_F/c_T = 17/7$, and $1/\theta = 0.1$, $T_L^* = 0.89$ is much less than $T_F^* = 3.66$.

When $L(T_A) \geq c_T/(c_F - c_T)$, $T_F^* \leq T_L^*$ and replacement first is better than replacement last, and conversely, when $L(T_A) < c_T/(c_F - c_T)$, $T_F^* > T_L^*$ and replacement last is better than replacement first. For example, when $1/\theta = 0.1$, $L(T_A) = 0.011$, and hence, $T_F^* = 0.12 < T_L^* = 0.13$ for $c_T/(c_F - c_T) = 0.01$, and $T_L^* = 0.18 < T_F^* = 0.24$ for $c_T/(c_F - c_T) = 0.03$.

Optimum T_F^* decreases to T_S^* with $1/\theta$ and T_L^* increases from T_S^* with $1/\theta$, because the unit is replaced before failure at time $\min\{T_F^*, Y\}$ for replacement first and at time $\max\{T_L^*, Y\}$ for replacement last. Furthermore, replacement first is better than replacement last as $1/\theta$ becomes larger. For example, when $c_T/(c_F - c_T) = 0.10$, if $1/\theta < 0.3$, then replacement last is better than replacement first, and if $1/\theta > 0.5$, then replacement first is better than replacement last. This means that if $1/\theta$ becomes larger, the unit has to be replaced mainly at a planned time T. So that, replacement first is better than replacement last, because the unit cannot be replaced before a random time Y for replacement last. It is of interest that when $1/\theta = 1.0$ and $c_T/(c_F - c_T) = 0.70$, $T_L^* = 0.91 < 1/\theta = 1.0 < T_F^* = 1.04$, and optimum replacement times are equal nearly to $1/\theta$. □

2.3.2 *Comparisons of Replacement Overtime, First, and Last*

Compare the expected cost rate $C_S(T)$ in (2.2) for standard replacement and $C_O(T)$ in (2.47) when $c_T = c_R$. It has been already shown that $T_O^* < T_S^*$ and when $h(\infty) = \infty$, both finite T_O^* and T_S^* exist. Furthermore, because T_S^* is an optimum solution of minimizing $C_S(T)$ in (2.2), $C_O(T_O^*)$ is greater than $C_S(T_S^*)$ from (2.49), i.e., standard replacement is better than replacement overtime. If $c_R < c_T$, then replacement overtime might be rather than standard replacement, as shown in Sect. 2.1. In this case, we would compute numerically $C_S(T_S^*)$ in (2.4) and $C_O(T_O^*)$ in (2.49), and compare them.

Example 2.5 (Comparison of replacements) We compute $\widehat{c}_{RO}$ in which both expected costs of standard replacement and replacement overtime are the same, when $c_R < c_T$, $G(t) = 1 - e^{-\theta t}$, and $F(t) = 1 - \exp(-t^2)$. First, compute T_S^* from (2.3), and the expected cost rate $C_S(T_S^*)$ from (2.4), where T_S^* have been already given in Table 2.2. Next, compute $\widehat{T}_O$ and $\widehat{c}_{RO}$ which satisfy the simultaneous equations:

$$(c_F - \widehat{c}_{RO})\widetilde{Q}_0(\widehat{T}_O; \theta) = (c_F - c_T)h(T_S^*),$$

$$\widetilde{Q}_0(\widehat{T}_O; \theta) \int_0^{\widehat{T}_O} \overline{F}(t)\, dt + \overline{F}(\widehat{T}_O) = \frac{c_F}{c_F - \widehat{c}_{RO}},$$

where

$$h(T) = 2T, \quad \widetilde{Q}_0(T; \theta) = \frac{\int_T^\infty e^{-\theta t} 2t e^{-t^2}\, dt}{\int_T^\infty e^{-\theta t} e^{-t^2}\, dt}.$$

That is, for given $c_T/(c_F - c_T)$ and T_S^*, we compute $\widehat{T}_O$ which satisfies

$$\int_0^{\widehat{T}_O} \overline{F}(t)\, dt + \frac{\overline{F}(\widehat{T}_O)}{\widetilde{Q}_0(\widehat{T}_O; \theta)} = \frac{c_F}{c_F - c_T} \frac{1}{h(T_S^*)},$$

and using $\widehat{T}_O$ and T_S^*, compute $\widehat{c}_{RO}$ which satisfies

$$(c_F - \widehat{c}_{RO})\widetilde{Q}_0(\widehat{T}_O; \theta) = (c_F - c_T)h(T_S^*).$$

Table 2.3 presents the replacement cost $\widehat{c}_{RO}$ when $C_S(T_S^*) = C_O(\widehat{T}_O)$. When $1/\theta = 0.5$, $1/\theta = 1.0$, and $c_T/(c_F - c_T) \leq 0.10$, there exists no $\widehat{c}_{RO}$, i.e., standard replacement is better than replacement overtime for any cost c_R. Table 2.3 indicates that $\widehat{c}_{RO}/c_T$ increases with $c_T/(c_F - c_T)$ from 0 to 1 and decreases with $1/\theta$. For

Table 2.3 Random replacement cost $\widehat{c}_{RO}$ when $C_S(T_S^*) = C_O(\widehat{T}_O)$

c_T	$1/\theta = 0.1$		$1/\theta = 0.5$		$1/\theta = 1.0$	
$c_F - c_T$	$\widehat{c}_{RO}/c_F$	$\widehat{c}_{RO}/c_T$	$\widehat{c}_{RO}/c_F$	$\widehat{c}_{RO}/c_T$	$\widehat{c}_{RO}/c_F$	$\widehat{c}_{RO}/c_T$
0.01	0.0002	0.0240	–	–	–	–
0.03	0.0198	0.6804	–	–	–	–
0.05	0.0386	0.8105	–	–	–	–
0.07	0.0567	0.8660	–	–	–	–
0.10	0.0825	0.9073	–	–	–	–
0.30	0.2240	0.9708	0.1445	0.6262	0.0303	0.1312
0.50	0.3277	0.9831	0.2665	0.7944	0.1833	0.5498
0.70	0.4070	0.9883	0.3579	0.8691	0.2941	0.7143
1.00	0.4961	0.9921	0.4592	0.9183	0.4139	0.8277

example, when $1/\theta = 0.1$, $\widehat{c}_{RO}$ becomes 0 as $c_T \to 0$ and becomes equal to c_T as $c_T \to c_F$ (Problem 8 in Sect. 2.5). □

Furthermore, when $c_T = c_R$ and $G(t) = 1 - e^{-\theta t}$, we compare replacement overtime with the other replacements: Comparing (2.26) with (2.48),

$$\widetilde{Q}_0(T;\theta)\int_0^T \overline{F}(t)\,dt - F(T) - h(T)\int_0^T e^{-\theta t}\overline{F}(t)\,dt + \int_0^T e^{-\theta t}dF(t)$$

$$= [\widetilde{Q}_0(T;\theta) - h(T)]\int_0^T \overline{F}(t)\,dt + \int_0^T (1 - e^{-\theta t})\overline{F}(t)[h(T) - h(t)]\,dt > 0,$$

and hence, $T_O^* < T_F^*$. In addition, from (2.27) and (2.49), and noting that $\widetilde{Q}_0(T;\theta) > h(T)$, if $\widetilde{Q}_0(T_O^*;\theta) < h(T_F^*)$, then replacement overtime is better than replacement first, and *vice versa*.

Comparing (2.39) with (2.48).

$$\widetilde{Q}_0(T;\theta)\int_0^T \overline{F}(t)\,dt - F(T) - h(T)\left[\int_0^T \overline{F}(t\,dt + \int_T^\infty e^{-\theta t}\overline{F}(t)\,dt\right]$$

$$+ 1 - \int_T^\infty (1 - e^{-\theta t})dF(t)$$

$$= [\widetilde{Q}_0(T;\theta) - h(T)]\int_0^T \overline{F}(t)\,dt + \int_T^\infty e^{-\theta t}\overline{F}(t)[h(t) - h(T)]\,dt > 0,$$

and hence, $T_O^* < T_L^*$. Thus, if $\widetilde{Q}_0(T_O^*; \theta) < h(T_L^*)$, then replacement overtime is better than replacement last, and *vice versa* (Problem 9 in Sect. 2.5).

For $0 < \theta < \infty$ and $0 < T < \infty$, we can obtain the following inequalities, using (11) of Appendix A.1 (Problem 10 in Sect. 2.5):

$$\frac{\int_T^\infty t\,dF(t)}{\int_T^\infty t\overline{F}(t)\,dt} > \frac{\int_T^\infty e^{-\theta t}\,dF(t)}{\int_T^\infty e^{-\theta t}\overline{F}(t)\,dt} > h(T) > \frac{\int_0^T t\,dF(t)}{\int_0^T t\overline{F}(t)\,dt} > \frac{\int_0^T e^{-\theta t}\,dF(t)}{\int_0^T e^{-\theta t}\overline{F}(t)\,dt}. \tag{2.53}$$

Compared (2.3) with (2.48) and (2.50) when $c_R = c_T$, $T_U^* > T_S^* > T_O^*$, and $C_U(T_U^*) > C_S(T_S^*)$ and $C_O(T_O^*) > C_S(T_S^*)$. Thus, compared (2.49) with (2.51), if $\widetilde{Q}_0(T_O^*; \theta) < Q_1(T_U^*)$ then replacement overtime is better than replacement with uniform random time, and *vice versa* (Problem 11 in Sect. 2.5).

2.4 Nth Working Time

It is assumed that Y_j $(j = 1, 2, \ldots)$ is the jth working time of a job in Fig. 2.2, and is independent and has an identical distribution $G(t) \equiv \Pr\{Y_j \le t\}$ with finite mean $1/\theta$ $(0 < \theta < \infty)$. That is, the unit works for a job with a renewal process according to a general distribution $G(t)$. Then, the probability that the unit works exactly j times in $[0, t]$ is $G^{(j)}(t) - G^{(j+1)}(t)$, where $G^{(j)}(t)$ $(j = 1, 2, \ldots)$ denotes the j-fold Stieltjes convolution of $G(t)$ with itself and $G^{(0)}(t) \equiv 1$ for $t \ge 0$. In addition, when $G^{(j)}(t)$ has a density function $g^{(j)}(t)$, i.e., $g^{(j)}(t) \equiv dG^{(j)}(t)/dt$, $r_j(t) \equiv g^{(j)}(t)/[1 - G^{(j)}(t)]$. Note that $r_j(t)dt$ represents the probability that the unit completes the jth work in $[t, t + dt]$, given that it operates for the jth number of working times at time t. For the above job with random working times, we take up the following two policies of replacement first and last.

2.4.1 Replacement First

Suppose that the unit is replaced before failure at a planned time T $(0 < T \le \infty)$ or at a planned number N $(N = 1, 2, \ldots)$ of working times, whichever occurs first. Then, the probability that the unit is replaced at time T is

$$\overline{F}(T)[1 - G^{(N)}(T)], \tag{2.54}$$

the probability that it is replaced at number N is

$$\int_0^T \overline{F}(t)\,dG^{(N)}(t), \tag{2.55}$$

and the probability that it is replaced at failure is

$$\int_0^T [1 - G^{(N)}(t)] dF(t),$$

(2.56)

where note that $(2.54) + (2.55) + (2.56) = 1$. Thus, the mean time to replacement is

$$T\overline{F}(T)[1 - G^{(N)}(T)] + \int_0^T t\overline{F}(t) dG^{(N)}(t) + \int_0^T t[1 - G^{(N)}(t)] dF(t)$$

$$= \int_0^T [1 - G^{(N)}(t)]\overline{F}(t) dt.$$

(2.57)

Therefore, the expected cost rate is

$$C_F(T, N) = \frac{c_T + (c_F - c_T) \int_0^T [1 - G^{(N)}(t)] dF(t) + (c_R - c_T) \int_0^T \overline{F}(t) dG^{(N)}(t)}{\int_0^T [1 - G^{(N)}(t)]\overline{F}(t) dt},$$

(2.58)

where c_F = replacement cost at failure, c_T = replacement cost at time T, and c_R = replacement cost at number N for any $N \geq 1$ with $c_F > c_T$ and $c_F > c_R$. By replacing $G(t)$ with $G^{(N)}(t)$ formally in (2.23), (2.58) is also obtained easily (Problem 12 in Sect. 2.5).

In particular, when the unit is replaced only at time T,

$$C_S(T) \equiv \lim_{N \to \infty} C_F(T, N) = \frac{c_F - (c_F - c_T)\overline{F}(T)}{\int_0^T \overline{F}(t) dt},$$

(2.59)

which agrees with (2.2), and when $N = 1$,

$$C_F(T) \equiv C_F(T, 1)$$

$$= \frac{c_T + (c_F - c_T) \int_0^T \overline{G}(t) dF(t) + (c_R - c_T) \int_0^T \overline{F}(t) dG(t)}{\int_0^T \overline{G}(t)\overline{F}(t) dt},$$

(2.60)

which agrees with (2.23).

Furthermore, when the unit is replaced only at number N,

$$C_F(N) \equiv \lim_{T \to \infty} C_F(T, N) = \frac{c_F - (c_F - c_R)\int_0^\infty \overline{F}(t)dG^{(N)}(t)}{\int_0^\infty [1 - G^{(N)}(t)]\overline{F}(t)\,dt} \quad (N = 1, 2, \ldots).$$

$$(2.61)$$

In particular, when each working time is constant at T_0, i.e., $G(t) \equiv 0$ for $t < T_0$, 1 for $t \geq T_0$, $G^{(N)}(t) = 0$ for $t < NT_0$, 1 for $t \geq NT_0$, and hence, the expected cost rate in (2.61) is

$$C_F(N) = \frac{c_F - (c_F - c_R)\overline{F}(NT_0)}{\int_0^{NT_0} \overline{F}(t)\,dt}, \qquad (2.62)$$

which agrees with the discrete age replacement in (9.1) of [8, p. 236].

2.4.1.1 Optimum N^*

We find an optimum N^* which minimizes $C_F(N)$ in (2.61). From the inequality $C_F(N+1) - C_F(N) \geq 0$,

$$Q_N \int_0^\infty [1 - G^{(N)}(t)]\overline{F}(t)\,dt - \int_0^\infty [1 - G^{(N)}(t)]dF(t) \geq \frac{c_R}{c_F - c_R}, \qquad (2.63)$$

where $Q_N \equiv \lim_{T \to \infty} Q_N(T)$ and for $0 < T \leq \infty$,

$$Q_N(T) \equiv \frac{\int_0^T [G^{(N)}(t) - G^{(N+1)}(t)]dF(t)}{\int_0^T [G^{(N)}(t) - G^{(N+1)}(t)]\overline{F}(t)\,dt}.$$

It is easily proved that if Q_N increases strictly to Q_∞, then the left-hand side of (2.63) also increases strictly to $\mu Q_\infty - 1$. Thus, if $Q_\infty > c_F/[\mu(c_F - c_R)]$, then there exists a finite and unique minimum N^* ($1 \leq N^* < \infty$) which satisfies (2.63).

In particular, when $G(t) = 1 - e^{-\theta t}$, i.e., $G^{(N)}(t) = \sum_{j=N}^\infty [(\theta t)^j/j!]e^{-\theta t}$, from **(3)** of Appendix A.1,

$$Q_N(T; \theta) = \frac{\int_0^T (\theta t)^N e^{-\theta t}\,dF(t)}{\int_0^T (\theta t)^N e^{-\theta t}\overline{F}(t)\,dt} \leq h(T) \qquad (2.64)$$

increases strictly with N to $h(T)$ and hence, Q_N increases strictly to $h(\infty)$. Therefore, if $h(\infty) > c_F/[\mu(c_F - c_R)]$, then there exists a finite and unique minimum N^* ($1 \leq N^* < \infty$) which satisfies (2.63).

2.4.1.2 Optimum T_F^* and N_F^*

We find optimum T_F^* and N_F^* which minimize $C_F(T, N)$ in (2.58). Differentiating $C_F(T, N)$ with respect to T and setting it equal to zero,

$$(c_F - c_T) \left\{ h(T) \int_0^T [1 - G^{(N)}(t)] \overline{F}(t) \, dt - \int_0^T [1 - G^{(N)}(t)] dF(t) \right\}$$

$$- (c_T - c_R) \left\{ r_N(T) \int_0^T [1 - G^{(N)}(t)] \overline{F}(t) \, dt - \int_0^T \overline{F}(t) dG^{(N)}(t) \right\} = c_T,$$

i.e.,

$$\int_0^T [1 - G^{(N)}(t)] \overline{F}(t) \{ (c_F - c_T)[h(T) - h(t)]$$

$$- (c_T - c_R)[r_N(T) - r_N(t)] \} \, dt = c_T, \qquad (2.65)$$

where $r_N(t) \equiv g^{(N)}(t)/[1 - G^{(N)}(t)]$ $(N = 1, 2, \ldots)$ and $r_0(t) \equiv 0$. From the inequality $C_F(T, N + 1) - C_F(T, N) \geq 0$,

$$\int_0^T [1 - G^{(N)}(t)] \overline{F}(t) \left((c_F - c_T)[Q_N(T) - h(t)] \right.$$

$$\left. + (c_T - c_R) \left\{ \frac{\int_0^T \overline{F}(t) d[G^{(N)}(t) - G^{(N+1)}(t)]}{\int_0^T [G^{(N)}(t) - G^{(N+1)}(t)] \overline{F}(t) \, dt} + r_N(t) \right\} \right) dt \geq c_T. \qquad (2.66)$$

In addition, substituting (2.65) for (2.66), (2.66) becomes

$$(c_F - c_T)[Q_N(T) - h(T)]$$

$$+ (c_T - c_R) \left\{ \frac{\int_0^T \overline{F}(t) d[G^{(N)}(t) - G^{(N+1)}(t)]}{\int_0^T [G^{(N)}(t) - G^{(N+1)}(t)] \overline{F}(t) \, dt} + r_N(T) \right\} \geq 0. \qquad (2.67)$$

Thus, when $c_T \leq c_R$, there does not exist finite optimum N_F^* for $T > 0$ because $Q_N(T) \leq h(T)$, i.e., $N_F^* = \infty$. In this case, the unit should be replaced only at time T.

Next, assume that $G(t) = 1 - e^{-\theta t}$, i.e., $G^{(N)}(t) = \sum_{j=N}^{\infty} [(\theta t)^j / j!] e^{-\theta t}$, and $c_T > c_R$. Then, (2.67) is

$$(c_F - c_T)[Q_N(T; \theta) - h(T)]$$

$$+ (c_T - c_R) \left[Q_N(T; \theta) + r_N(T) + \frac{(\theta T)^N e^{-\theta T} \overline{F}(T)}{\int_0^T (\theta t)^N e^{-\theta t} \overline{F}(t)\, dt} \right] \geq 0, \qquad (2.68)$$

where $Q_N(T; \theta)$ is given in (2.64) and

$$r_N(T) = \frac{\theta(\theta T)^{N-1}/(N-1)!}{\sum_{j=0}^{N-1} [(\theta T)^j / j!]} \quad (N = 1, 2, \ldots).$$

Recalling that $Q_N(T; \theta)$ increases with N to $h(T)$ and $r_N(T)$ decreases with N to 0 from (1) and (3) of Appendix A.1, there exists a finite N_F^* ($1 \leq N_F^* < \infty$) which satisfies (2.68) for $T > 0$.

Furthermore, when $G(t) = 1 - e^{-\theta t}$, the left-hand side of (2.65) goes to

$$\int_0^\infty [1 - G^{(N)}(t)] \overline{F}(t)\{(c_F - c_T)[h(\infty) - h(t)] - (c_T - c_R)[\theta - r_N(t)]\}\, dt, \qquad (2.69)$$

as $T \to \infty$, because $r_N(T)$ increases with T to θ. Therefore, if (2.69) is greater than c_T, then there exits a finite T_F^* ($0 < T_F^* < \infty$) which satisfies (2.65). It can be clearly seen that if $h(\infty) = \infty$, then (2.69) becomes ∞. In this case, the resulting cost rate is

$$C_F(T_F^*, N_F^*) = (c_F - c_T)h(T_F^*) - (c_T - c_R)r_{N_F^*}(T_F^*). \qquad (2.70)$$

When $N_F^* = \infty$, $r_\infty(T) = 0$ and the expected cost rate is given in (2.27).

Example 2.6 (*Replacement for Weibull failure time*) Table 2.4 presents optimum N_F^* which minimizes $C_F(N)$ in (2.61), and T_F^* and N_F^* which minimize $C_F(T, N)$ in (2.58) when $G(t) = 1 - e^{-t}$ and $F(t) = 1 - \exp[-(t/10)^2]$. This indicates that optimum N_F^* given in (2.63) is constant for different c_T. Optimum T_F^* given in (2.65) increases, and N_F^* given in (2.68) decreases with c_T, however, is almost constant except for $N_F^* = \infty$, and becomes equal to N_F^* in (2.63) as c_T becomes larger.   □

2.4.2 Replacement Last

Suppose that the unit is replaced before failure at time T ($0 \leq T < \infty$) or at number N ($N = 0, 1, 2, \ldots$), whichever occurs last. Then, the probability that the unit is replaced at time T is

$$\overline{F}(T)G^{(N)}(T), \qquad (2.71)$$

the probability that it is replaced at number N is

Table 2.4 Optimum T_F^* and N_F^* when $G(t) = 1 - e^{-t}$, $F(t) = 1 - e^{-(t/10)^2}$, and $c_F = 100$

c_T	$c_R = 10$					$c_R = 15$				
	N_F^*	$C_F(N_F^*)$	T_F^*	N_F^*	$C_F(T_F^*, N_F^*)$	N_F^*	$C_F(N_F^*)$	T_F^*	N_F^*	$C_F(T_F^*, N_F^*)$
10	4	24.94	3.36	∞	6.06	5	34.98	3.36	∞	6.06
11	4	24.94	3.64	6	6.31	5	34.98	3.55	∞	6.32
12	4	24.94	4.05	5	6.50	5	34.98	3.74	∞	6.57
13	4	24.94	4.68	4	6.62	5	34.98	3.91	∞	6.81
14	4	24.94	5.14	4	6.69	5	34.98	4.09	∞	7.04
15	4	24.94	5.64	4	6.73	5	34.98	4.26	∞	7.25
16	4	24.94	6.17	4	6.77	5	34.98	4.49	8	7.45
17	4	24.94	6.73	4	6.79	5	34.98	4.92	6	7.61
18	4	24.94	7.32	4	6.80	5	34.98	5.26	6	7.72
19	4	24.94	7.95	4	6.81	5	34.98	5.96	5	7.79
20	4	24.94	8.60	4	6.82	5	34.98	6.45	5	7.83

$$\int_T^\infty \overline{F}(t) dG^{(N)}(t),$$ (2.72)

and the probability that it is replaced at failure is

$$F(T) + \int_T^\infty [1 - G^{(N)}(t)] dF(t) = 1 - \int_T^\infty G^{(N)}(t) dF(t),$$ (2.73)

where note that $(2.71) + (2.72) + (2.73) = 1$. Thus, the mean time to replacement is

$$T\overline{F}(T)G^{(N)}(T) + \int_T^\infty t\overline{F}(t) dG^{(N)}(t) + \int_0^T t dF(t) + \int_T^\infty t[1 - G^{(N)}(t)] dF(t)$$

$$= \int_0^T \overline{F}(t) dt + \int_T^\infty \overline{F}(t)[1 - G^{(N)}(t)] dt.$$ (2.74)

Therefore, the expected cost rate is

$$C_L(T, N) = \frac{c_F - (c_F - c_T) \int_T^\infty G^{(N)}(t) dF(t) + (c_R - c_T) \int_T^\infty \overline{F}(t) dG^{(N)}(t)}{\mu - \int_T^\infty \overline{F}(t) G^{(N)}(t) dt}.$$ (2.75)

By replacing $G(t)$ with $G^{(N)}(t)$ formally in (2.37), (2.75) is also obtained easily (Problem 13 in Sect. 2.5). Compared to $C_F(T, N)$ in (2.58) when $c_T = c_R$, both numerator and denominator are larger than those in (2.58). In particular, when $N = 0$,

$C_L(T, 0)$ agrees with $C_S(T)$ in (2.24), and when $T = 0$, $C_L(0, N)$ agrees with $C_F(N)$ in (2.61).

We find optimum T_L^* and N_L^* which minimize $C_L(T, N)$ in (2.75). Differentiating $C_L(T, N)$ with respect to T and setting it equal to zero,

$$
(c_F - c_T) \left\{ h(T) \left[\mu - \int_T^\infty \overline{F}(t) G^{(N)}(t) \, dt \right] \right.
$$
$$
\left. + \int_T^\infty G^{(N)}(t) dF(t) \right\} + (c_T - c_R) \left\{ \tilde{r}_N(T) \left[\mu - \int_T^\infty \overline{F}(t) G^{(N)}(t) \, dt \right] \right.
$$
$$
\left. + \int_T^\infty \overline{F}(t) dG^{(N)}(t) \right\} = c_F, \tag{2.76}
$$

where $\tilde{r}_N(T) \equiv g^{(N)}(T)/G^{(N)}(T)$. From the inequality $C_L(T, N + 1) - C_L(T, N) \geq 0$,

$$
(c_F - c_T) \left\{ \tilde{Q}_N(T) \left[\mu - \int_T^\infty \overline{F}(t) G^{(N)}(t) \, dt \right] \right.
$$
$$
\left. + \int_T^\infty G^{(N)}(t) dF(t) \right\} + (c_T - c_R) \left\{ \tilde{Q}_N(T) \left[\mu - \int_T^\infty \overline{F}(t) G^{(N)}(t) \, dt \right] \right.
$$
$$
\left. + \int_T^\infty \overline{F}(t) dG^{(N)}(t) \right\} \geq c_F. \tag{2.77}
$$

Furthermore, substituting (2.76) for (2.77), (2.77) becomes

$$
(c_F - c_T)[\tilde{Q}_N(T) - h(T)] + (c_T - c_R)[\tilde{Q}_N(T) - \tilde{r}_N(T)] \geq 0, \tag{2.78}
$$

where

$$
\tilde{Q}_N(T) \equiv \frac{\int_T^\infty [G^{(N)}(t) - G^{(N+1)}(t)] dF(t)}{\int_T^\infty [G^{(N)}(t) - G^{(N+1)}(t)] \overline{F}(t) \, dt} \geq h(T).
$$

Thus, if $c_R \geq c_T$ and

$$
\tilde{r}_N(T) \geq \tilde{Q}_N(T), \tag{2.79}
$$

then $C_L(T, N + 1) - C_L(T, N) \geq 0$ for all $N \geq 0$ and $T > 0$, i.e., $N_L^* = 0$. In this case, the unit should be replaced only at time T.

Example 2.7 (Replacement for Weibull failure time) Table 2.5 presents optimum T_L^* and N_L^* which minimize $C_L(T, N)$ in (2.75) when $G(t) = 1 - e^{-t}$ and $F(t) = 1 - \exp[-(t/10)^2]$.

Table 2.5 Optimum T_L^* and N_L^* when $G(t) = 1 - e^{-t}$, $F(t) = 1 - e^{-(t/10)^2}$, and $c_F = 100$

c_T	$c_R = 10$			$c_R = 15$		
	T_L^*	N_L^*	$C_L(T_L^*, N_L^*)$	T_L^*	N_L^*	$C_L(T_L^*, N_L^*)$
10	3.36	0	6.06	3.37	0	6.07
11	3.55	0	6.32	3.56	0	6.33
12	3.70	1	6.57	3.73	0	6.58
13	3.72	2	6.79	3.92	0	6.81
14	0	4	6.82	4.09	0	7.04
15	0	4	6.82	4.26	0	7.25
16	0	4	6.82	4.44	0	7.45
17	0	4	6.82	4.59	1	7.64
18	0	4	6.82	4.68	2	7.82
19	0	4	6.82	0	5	7.91
20	0	4	6.82	0	5	7.91

This indicates that optimum T_L^* given in (2.76) and N_L^* given in (2.78) increase with c_T except $T_L^* = 0$, N_L^* is constant for $T_L^* = 0$. Clearly, $N_L^* = 0$ for $c_R \geq c_T$. Compared to Table 2.4, the expected cost $C_L(T_L^*, N_L^*)$ is a little greater than $C_F(T_F^*, N_F^*)$. □

2.4.3 Replacement with Constant Time

Suppose that when $c_T = c_R < c_F$, a planned replacement time T $(0 < T < \infty)$ is fixed. It would be estimated from the above discussions that if $T \leq T_S^*$ then the unit should be replaced only at time T. From (2.58), we find an optimum N_F^* of replacement first for a given T, which minimizes the expected cost rate

$$C_F(N; T) = \frac{c_T + (c_F - c_T) \int_0^T [1 - G^{(N)}(t)] \mathrm{d}F(t)}{\int_0^T [1 - G^{(N)}(t)] \overline{F}(t) \, \mathrm{d}t} \qquad (N = 1, 2, \ldots). \qquad (2.80)$$

From the inequality $C_F(N + 1; T) - C_F(N; T) \geq 0$, i.e., setting by $c_T = c_R$ in (2.66),

$$Q_N(T) \int_0^T [1 - G^{(N)}(t)] \overline{F}(t) \, \mathrm{d}t - \int_0^T [1 - G^{(N)}(t)] \mathrm{d}F(t) \geq \frac{c_T}{c_F - c_T}, \qquad (2.81)$$

where $Q_N(T)$ is given in (2.63). From (2.3), (2.81) is rewritten as

$$Q_N(T) \int_0^T [1 - G^{(N)}(t)]\overline{F}(t)\,dt - \int_0^T [1 - G^{(N)}(t)]dF(t)$$

$$\geq h(T_S^*) \int_0^{T_S^*} \overline{F}(t)\,dt - F(T_S^*). \tag{2.82}$$

Recalling that when $G(t) = 1 - e^{-\theta t}$, $Q_N(T; \theta)$ in (2.64) increases strictly with N to $h(T)$, the left-hand side of (2.82) also increases strictly with N to

$$h(T) \int_0^T \overline{F}(t)\,dt - F(T).$$

Thus, if $T \leq T_S^*$, then there does not exist any N which satisfies (2.82), i.e., $N_F^* = \infty$. Conversely, if $T > T_S^*$, then there exists a finite and unique minimum N_F^* which satisfies (2.81) when $G(t) = 1 - e^{-\theta t}$.

Next, from (2.75), we find an optimum N_L^* of replacement last for a given T which minimizes the expected cost rate

$$C_L(N; T) = \frac{c_F - (c_F - c_T) \int_T^\infty G^{(N)}(t)dF(t)}{\mu - \int_T^\infty \overline{F}(t)G^{(N)}(t)\,dt} \quad (N = 0, 1, 2, \ldots). \tag{2.83}$$

From the inequality $C_L(N + 1; T) - C_L(N; T) \geq 0$, i.e., setting by $c_T = c_R$ in (2.77),

$$h(T) \left[\mu - \int_T^\infty \overline{F}(t)G^{(N)}(t)\,dt \right] + \int_T^\infty G^{(N)}(t)dF(t) \geq \frac{c_F}{c_F - c_T}. \tag{2.84}$$

From (2.3) to (2.84) is rewritten as

$$h(T) \left[\mu - \int_T^\infty \overline{F}(t)G^{(N)}(t)\,dt \right] + \int_T^\infty G^{(N)}(t)dF(t)$$

$$\geq h(T_S^*) \int_0^{T_S^*} \overline{F}(t)\,dt + \overline{F}(T_S^*). \tag{2.85}$$

When $G(t) = 1 - e^{-\theta t}$, the left-hand side of (2.85) increases strictly with N from $\tilde{Q}_0(T; \theta) \int_0^T \overline{F}(t)dt + \overline{F}(T)$ to $\mu h(\infty)$ from (6) of Appendix A.1. Thus, if $T \geq T_S^*$ then $N_L^* = 0$, i.e., the unit is replaced only at time T. Conversely, if $T < T_S^*$ and $h(\infty) = \infty$, then there exists a finite and unique minimum N_L^* which satisfies (2.85) (Problem 14 in Sect. 2.5). This means that if $T > T_S^*$ then we should adopt replacement first, and conversely, if $T < T_S^*$ then we should adopt replacement last.

2.5 Problems

1 Show that $Q(t) \equiv c_1 F(t) + c_R \overline{F}(t)$ and $S(t) \equiv \int_0^t \overline{F}(u)du$.
2 Derive (2.9) and (2.11).
3 Prove that $\theta_1^* > \theta_2^*$.
4 Prove that T_F^* increases with θ from T_S^*.
5 Prove that T_L^* decreases with θ to T_S^*.
6 Prove that $(2.42) + (2.43) + (2.44) = 1$.
7 Show that if $h(\infty) > c_F/[\mu(c_F - c_R)]$ then there exists a finite and unique T_O^* which satisfies (2.48), and $C_O(T_O^*)$ is given in (2.49).
8 Compute $\hat{c}_{RO}/c_F$ and $\hat{c}_{RO}/c_T$ in Table 2.3.
9 Show that if $\tilde{Q}_0(T_O^*; \theta) < h(T_F^*)$ and $\tilde{Q}_0(T_O^*; \theta) < h(T_L^*)$, then replacement overtime is better than both replacement first and last.
10 Prove that

$$\frac{\int_0^T t^\alpha dF(t)}{\int_0^T t^\alpha \overline{F}(t)\,dt} \quad \text{and} \quad \frac{\int_T^\infty t^\alpha dF(t)}{\int_T^\infty t^\alpha d\overline{F}(t)dt} \tag{2.86}$$

increases with α $(0 < \alpha < \infty)$.
11 Give numerical examples of $C_O(T_O^*)$ and $C_U(T_U^*)$, and compare them.
*12 When the unit is replaced at number N before time T, and it is replaced at the first completion of working time over time T before the Nth time, obtain the expected cost rate and discuss an optimum policy.
*13 When the unit is replaced at time T after the Nth working time, and it is replaced at the first completion of working time over time T before the Nth time, obtain the expected cost rate and discuss an optimum policy.
14 Make a numerical example of optimum N_F^* and N_L^* for a given T when $F(t) = 1 - \exp(-t^2)$.

References

1. Barlow RE, Proschan F (1965) Mathematical theory of reliability. Wiley, New York
2. Chen M, Mizutani S, Nakagawa T (2010) Random and age replacement policies. Inter J Reliab Qual Saf Eng 17:27–39
3. Nakagawa T, Zhao X, Yun WY (2011) Optimal age replacement and inspection policies with random failure and replacement times. Inter J Reliab Qual Saf Eng 18:1–12
4. Zhao X, Nakagawa T (2012) Optimization problems of replacement first or last in reliability theory. Euro J Oper Res 223:141–149
5. Zhao X, Qian C, Nakamura T (2014) Optimal age and periodic replacement with overtime policies. To appear in Inter J Reliab Qual Saf Eng 21
6. Ross SM (1983) Stochastic processes. Wiley, New York
7. Nakagawa T (2011) Stochastic processes with applications to reliability theory. Springer, London
8. Nakagawa T (2005) Maintenance theory of reliability. Springer, London
9. Nakagawa T (2008) Advanced reliability models and maintenance policies. Springer, London

Chapter 3
Random Periodic Replacement Policies

When large and complex systems consist of many kinds of units, we should make minimal repair at each failure, and make the planned replacement or preventive maintenance at periodic times. This is called the periodic replacement with minimal repair at failures and was summarized [1, p. 96], [2, p. 95]. A large number of mathematical models, which can be used in making minimal repair and maintenance plans efficiently, were introduced [3].

The unit works for a job with random working times [2, p. 245]: The age and periodic replacement policies for random working times were derived analytically [4, 5]. It has been assumed in all policies that the unit is replaced at a planned time or at the completion of working times, whichever occurs first. Such policies would be reasonable in practical fields if the replacement cost after failure might be much high. However, if this cost would be estimated to be not so high, then the unit should be working as long as possible. From such viewpoints, the replacement policies, where the unit is replaced at a planned time or a working time, whichever occurs last, were proposed, and the expected cost rates were obtained [4, 5]. The various schedules of jobs which have random working and processing times were summarized [6].

It has been already wellknown that an optimum maintenance is nonrandom for an infinite time span [1, p. 86]. It may not be true in the case where units work for a job with random working and processing times, because replacements during its working time are impossible or impractical [1, p. 72], and replacement costs after the completion of working times might be lower than those at planned times. From these viewpoints, several random age replacement policies have been proposed, and their optimum policies which minimize the expected cost rates have been discussed analytically and numerically in Chap. 2.

By similar considerations and methods to random age replacement policies in Chap. 2, we make minimal repair at failures and take up the following four replacement policies:

1. Standard replacement: The unit is replaced at periodic times kT ($k = 1, 2, \ldots$).
2. Replacement first: The unit is replaced at time T or at a working time Y, whichever occurs first.

© Springer-Verlag London 2014
T. Nakagawa, *Random Maintenance Policies*,
Springer Series in Reliability Engineering, DOI 10.1007/978-1-4471-6575-0_3

3. Replacement last: The unit is replaced at time T or at a working time Y, whichever occurs last.
4. Replacement overtime: The unit is replaced at the first completion of working times over time T.

Policies 2 and 3 are the extended ones of standard replacement and Policy 4 is a modified standard replacement. First, we show that an optimum policy for random periodic replacement is nonrandom, as shown in age replacement. In addition, an optimum policy for random block replacement is also shown to be nonrandom.

We summarize optimum replacement times for four replacement policies. Next, we compare each policy with one another when the working time is exponential and the replacement costs are the same. For example, it is shown theoretically that standard replacement is the best among four ones, as estimated previously. Furthermore, we determine either replacement first or last is better than the other according to the ratio of replacement cost to minimal repair cost. Finally, we discuss theoretically and numerically that if how much the cost of random replacement is lower than that of periodic replacement, then random replacements are the same as standard one.

Furthermore, we consider replacement policies with two variables [7, p. 149], where the unit is replaced at a planned time T or at the Nth completion of working times, whichever occurs first and last. Two expected cost rates are obtained, and optimum policies which minimize them are derived analytically. Finally, we take up two policies of replacement overtime where the unit is replaced at the Nth number of working times or at the first completion of working times over time T, whichever occurs first and last, and obtain their expected cost rates.

Throughout this chapter, suppose that the unit has a failure distribution $F(t)$ with finite mean $\mu \equiv \int_0^\infty \overline{F}(t)\,dt < \infty$, where $\overline{\Phi}(t) \equiv 1 - \Phi(t)$. When $F(t)$ has a density function $f(t)$, i.e., $f(t) \equiv dF(t)/dt$, the failure rate is $h(t) \equiv f(t)/\overline{F}(t)$ for $F(t) < 1$ and increases strictly from $h(0) = 0$ to $h(\infty) \equiv \lim_{t\to\infty} h(t)$, which might be infinity. In addition, the cumulative hazard rate is $H(t) \equiv \int_0^t h(u)\,du$, a renewal function is $M(t) \equiv \sum_{j=1}^\infty F^{(j)}(t)$, and a renewal density is $m(t) \equiv dM(t)/dt$, where $\Phi^{(j)}(t)$ is the j-fold Stieltjes convolution of $\Phi(t)$ with itself ($j = 1, 2, \ldots$) and $\Phi^{(0)}(t) \equiv 1$ for $t \geq 0$.

3.1 Random Replacement

A unit is replaced at time Y and undergoes minimal repair at each failure between replacements, where Y is a random variable with a general distribution $G(t)$ with finite mean. Then, because failures occur at a nonhomogeneous Poisson process with mean value function $H(t)$ [2, p. 98], [8, p. 27], the expected number of failures in $[0, t]$ is $H(t)$. Thus, the expected number of failures between replacements is

$$\int_0^\infty H(t)\,dG(t) = \int_0^\infty \overline{G}(t)h(t)\,dt,$$

and the mean time to replacement is

$$\int_0^\infty t\,dG(t) = \int_0^\infty \overline{G}(t)\,dt.$$

Therefore, the expected cost rate is [2, p.247]

$$C_P(G) = \frac{c_M \int_0^\infty H(t)\,dG(t) + c_R}{\int_0^\infty \overline{G}(t)\,dt},\qquad (3.1)$$

where c_M = minimal repair cost at each failure and c_R = replacement cost at random time Y.

We can write (3.1) as

$$C_P(G) = \frac{\int_0^\infty Q(t)\,dG(t)}{\int_0^\infty S(t)\,dG(t)},$$

where

$$Q(t) \equiv c_M H(t) + c_R, \quad S(t) \equiv t.$$

Suppose that there exists a minimum value T $(0 < T \le \infty)$ of $Q(t)/S(t)$. Because

$$\frac{Q(t)}{S(t)} \ge \frac{Q(T)}{S(T)},$$

it follows that

$$\int_0^\infty Q(t)\,dG(t) \ge \frac{Q(T)}{S(T)} \int_0^\infty S(t)\,dG(t).$$

So that,

$$C_P(G) \ge \frac{Q(T)}{S(T)} = C_P(G_T),$$

where $G_T(t)$ is the degenerate distribution placing unit mass at T, i.e., $G_T(t) \equiv 1$ for $t \ge T$ and $G_T(t) \equiv 0$ for $t < T$. If $T = \infty$ then the unit is not replaced and undergoes always only minimal repair at each failure.

Thus, an optimum replacement is nonrandom, and the expected cost rate is

$$C_S(T) = \frac{c_M H(T) + c_T}{T},\qquad (3.2)$$

where c_T = replacement cost at time T. In particular, when the unit is not replaced and undergoes only minimal repair at each failure, the expected cost rate is

$$C_S(\infty) \equiv \lim_{T \to \infty} C_S(T) = c_M h(\infty). \tag{3.3}$$

If $\int_0^\infty t \, dh(t) > c_T/c_M$, then an optimum time T_S^* $(0 < T_S^* < \infty)$ which minimizes $C_S(T)$ satisfies (Problem 1 in Sect. 3.4)

$$Th(T) - H(T) = \frac{c_T}{c_M}, \quad \int_0^T t \, dh(t) = \frac{c_T}{c_M},$$

or

$$\int_0^T [h(T) - h(t)] \, dt = \frac{c_T}{c_M}, \tag{3.4}$$

and the resulting cost rate is

$$C_S(T_S^*) = c_M h(T_S^*). \tag{3.5}$$

Example 3.1 (Periodic replacement for exponential random and Weibull failure times) When $G(t) = 1 - e^{-\theta t}$ $(0 < \theta < \infty)$ and $H(t) = \lambda t^\alpha$ $(\alpha > 1)$, we find an optimum θ^* which minimizes the expected cost rate in (3.1) given by

$$C_P(\theta) = \frac{c_M \int_0^\infty e^{-\theta t} \lambda \alpha t^{\alpha-1} dt + c_R}{1/\theta} = c_M \lambda \Gamma(\alpha + 1)\theta^{1-\alpha} + c_R \theta, \tag{3.6}$$

where $\Gamma(\alpha) \equiv \int_0^\infty x^{\alpha-1} e^{-x} \, dx$ for $\alpha > 0$. An optimum θ^* which minimizes $C_P(\theta)$ is easily given by

$$\frac{1}{\theta^*} = \left[\frac{c_R}{c_M} \frac{1}{\lambda(\alpha - 1)\Gamma(\alpha + 1)} \right]^{1/\alpha}, \tag{3.7}$$

and the resulting cost rate is

$$C_P(\theta^*) = \frac{c_M \lambda \alpha \Gamma(\alpha + 1)}{(\theta^*)^{\alpha-1}}. \tag{3.8}$$

On the other hand, an optimum T_S^* which satisfies (3.4) is

$$T_S^* = \left[\frac{c_T}{c_M} \frac{1}{\lambda(\alpha - 1)} \right]^{1/\alpha}, \tag{3.9}$$

Table 3.1 Optimum T_S^*, $1/\theta^*$, $1/\hat{\theta}$, $\hat{c}_R/c_M$, and $\hat{c}_R/c_T$ when $F(t) = 1 - e^{-t^\alpha}$

c_T/c_M or c_R/c_M	$\alpha = 2$					$\alpha = 3$				
	T_S^*	$1/\theta^*$	$1/\hat{\theta}$	$\hat{c}_R/c_M$	$\hat{c}_R/c_T$	T_S^*	$1/\theta^*$	$1/\hat{\theta}$	$\hat{c}_R/c_M$	$\hat{c}_R/c_T$
0.1	0.316	0.224	0.158	0.050	0.500	0.368	0.203	0.150	0.041	0.410
0.2	0.477	0.316	0.224	0.100	0.500	0.464	0.255	0.189	0.082	0.410
0.5	0.707	0.500	0.354	0.250	0.500	0.630	0.347	0.257	0.204	0.410
1.0	1.000	0.707	0.500	0.500	0.500	0.794	0.437	0.324	0.408	0.410
2.0	1.414	1.000	0.707	1.000	0.500	1.000	0.550	0.408	0.816	0.410
5.0	2.236	1.581	1.118	2.500	0.500	1.357	0.747	0.554	2.041	0.410
10.0	3.162	2.236	1.581	5.000	0.500	1.710	0.941	0.698	4.082	0.410

and the resulting cost rate is

$$C_S(T_S^*) = c_M \lambda \alpha (T_S^*)^{\alpha - 1}. \tag{3.10}$$

It can be easily seen that when $c_T = c_R$, $T_S^* = [\Gamma(\alpha + 1)]^{1/\alpha}/\theta^*$, and hence, $1/\theta^* < T_S^*$ and $C_S(T_S^*) < C_P(\theta^*)$. So that, standard replacement is better than random one, as already shown in Sect. 2.1. Furthermore, when $c_T > c_R$, we compute a replacement cost $\hat{c}_R$ in which both expected cost rates $C_P(\theta^*)$ in (3.8) and $C_S(T_S^*)$ in (3.10) are the same. From $C_S(T_S^*) = C_P(\hat{\theta})$,

$$\frac{1}{\hat{\theta}} = \frac{T_S^*}{[\Gamma(\alpha + 1)]^{1/(\alpha - 1)}}.$$

Using $\hat{\theta}$, we compute

$$\frac{\hat{c}_R}{c_M} = \frac{\lambda(\alpha - 1)\Gamma(\alpha + 1)}{(\hat{\theta})^\alpha} = \frac{c_T}{c_M} \frac{1}{[\Gamma(\alpha + 1)]^{1/(\alpha - 1)}}. \tag{3.11}$$

Table 3.1 presents optimum T_S^*, $1/\theta^*$, $1/\hat{\theta}$, $\hat{c}_R/c_M$, and $\hat{c}_R/c_T$ for c_T/c_M or c_R/c_M, when $F(t) = 1 - \exp(-t^\alpha)$ ($\alpha = 2, 3$). This indicates that $T_S^* > 1/\theta^* > 1/\hat{\theta}$ and $\hat{c}_R/c_M$ decreases with α and increases with c_T/c_M or c_R/c_M. It is of great interest that $\hat{c}_R/c_T$ depends only on α, because from (3.11), $\hat{c}_R/c_T = \Gamma(\alpha + 1)^{-1/(\alpha - 1)}$. For example, when $\alpha = 2$, $\hat{c}_R/c_T = 0.5$, i.e., when the random replacement cost is 50% of the periodic one, both expected costs $C_S(T_S^*)$ and $C_P(\theta^*)$ are the same. □

Example 3.2 (Replacement for uniform random time) Suppose that Y has a uniform distribution for the interval $[0, T]$ ($0 < T < \infty$), i.e., $G(t) = t/T$ for $t \leq T$ and 1 for $t > T$ in Example 2.2. Then, the expected cost rate in (3.1) is a function of T and is given by

$$C_U(T) = \frac{c_M \int_0^T (T - t)h(t)\,dt + c_R T}{T^2/2}. \tag{3.12}$$

Clearly,

$$C_U(0) \equiv \lim_{T \to 0} C_U(T) = \infty, \quad C_U(\infty) \equiv \lim_{T \to \infty} C_U(T) = c_M h(\infty),$$

which agrees with (3.3). We find an optimum T_U^* $(0 < T_U^* \le \infty)$ which minimizes $C_U(T)$. Differentiating $C_U(T)$ with respect to T and setting it equal to zero,

$$\int_0^T \left(\frac{2t}{T} - 1 \right) h(t) \, dt = \frac{c_R}{c_M}, \tag{3.13}$$

whose left-hand side increases strictly with T from 0. Furthermore,

$$\lim_{T \to \infty} \frac{2 \int_0^T t h(t) \, dt - T H(T)}{T} = \lim_{T \to \infty} [Th(T) - H(T)]$$

$$= \lim_{T \to \infty} \int_0^T t \, dh(t) = \int_0^\infty t \, dh(t).$$

Thus, if $\int_0^\infty t \, dh(t) > c_R / c_M$, then there exists a finite and unique T_U^* $(0 < T_U^* < \infty)$ which satisfies (3.13), and the resulting cost rate is

$$C_U(T_U^*) = \frac{c_M H(T_U^*) + c_R}{T_U^*}. \tag{3.14}$$

In the case of $c_T = c_R$, $T_U^* > T_S^*$ and $C_U(T_U^*) = C_S(T_U^*)$ in (3.2). Thus, the replacement with uniform random time is not better than standard replacement. In particular, when $H(t) = \lambda t^\alpha$ $(\alpha > 1)$, (3.13) becomes

$$\lambda \left(\frac{\alpha - 1}{\alpha + 1} \right) (T_U^*)^\alpha = \frac{c_R}{c_M}, \quad \text{i.e.,} \quad T_U^* = \left(\frac{c_R}{c_M} \frac{\alpha + 1}{\lambda(\alpha - 1)} \right)^{1/\alpha},$$

and when $c_T = c_R$, $T_U^* > T_S^*$. $\square$

3.1.1 Block Replacement

A unit is replaced at time Y, and is replaced with a new one at each failure until time Y. Then, the expected number of failures in $[0, t]$ is $M(t)$ [2, p. 118], [8, p. 61]. Thus, replacing $H(t)$ in (3.1) with $M(t)$, the expected cost rate is

$$C_B(G) = \frac{c_F \int_0^\infty M(t)\, dG(t) + c_R}{\int_0^\infty \overline{G}(t)\, dt},$$ (3.15)

where c_F = replacement cost at each failure and c_R is given in (3.1).

By similar arguments to periodic replacement (Problem 2 in Sect. 3.4), an optimum replacement is nonrandom, and the expected cost rate with replacement time T $(0 < T \le \infty)$ is [2, p. 117]

$$C_B(T) = \frac{c_F M(T) + c_T}{T},$$ (3.16)

where c_T = replacement cost at time T. If $\int_0^\infty t\, dm(t) > c_T/c_F$, then an optimum T_B^* which minimizes $C_B(T)$ satisfies

$$Tm(T) - M(T) = \frac{c_T}{c_F},$$ (3.17)

and the resulting cost rate is

$$C_B(T_B^*) = c_F m(T_B^*).$$ (3.18)

Example 3.3 (Block replacement for gamma failure time and exponential random time) When $G(t) = 1 - e^{-\theta t}$, the expected cost rate in (3.15) is

$$C_B(\theta) = \theta[c_F M^*(\theta) + c_R],$$ (3.19)

where $\Phi^*(s)$ is the Laplace-Stieltjes (LS) transform of $\Phi(t)$, i.e., $\Phi^*(s) \equiv \int_0^\infty e^{-st}\, d\Phi(t)$ for $Re(s) > 0$. In particular, when the failure time has a gamma density function $f(t) = [\lambda(\lambda t)^{\alpha-1}/\Gamma(\alpha)]e^{-\lambda t}$ for $\alpha > 1$,

$$M^*(\theta) = \frac{F^*(\theta)}{1 - F^*(\theta)} = \frac{\lambda^\alpha}{(\theta + \lambda)^\alpha - \lambda^\alpha},$$

and hence, the expected cost rate is

$$C_B(\theta) = \theta \left[\frac{c_F \lambda^\alpha}{(\theta + \lambda)^\alpha - \lambda^\alpha} + c_R \right].$$ (3.20)

We find an optimum θ^* which minimizes $C_B(\theta)$. Differentiating $C_B(\theta)$ with θ and setting it equal to zero,

$$\frac{\lambda^\alpha}{(\theta + \lambda)^\alpha - \lambda^\alpha} \left[\frac{\alpha\theta(\theta + \lambda)^{\alpha-1}}{(\theta + \lambda)^\alpha - \lambda^\alpha} - 1 \right] = \frac{c_R}{c_F}.$$ (3.21)

Table 3.2 Optimum T_B^* and $1/\theta^*$ when $f(t) = [t^{\alpha-1}/(\alpha-1)!]e^{-t}$

c_T/c_F or c_R/c_F	$\alpha = 2$		$\alpha = 3$		$\alpha = 4$	
	T_B^*	$1/\theta^*$	T_B^*	$1/\theta^*$	T_B^*	$1/\theta^*$
0.01	0.157	0.125	0.355	0.234	0.630	0.365
0.05	0.412	0.404	0.691	0.565	1.059	0.770
0.10	0.688	0.861	0.969	0.980	1.374	1.234
0.15	1.011	1.716	1.222	1.553	1.634	1.814
0.20	1.497	4.236	1.487	2.500	1.881	2.667
0.25	∞	∞	1.801	4.536	2.135	4.144
0.30	∞	∞	2.255	12.565	2.419	7.524
0.35	∞	∞	∞	∞	2.777	24.232
0.40	∞	∞	∞	∞	∞	∞

For example, when $\alpha = 2$,

$$\frac{\lambda}{\theta + 2\lambda} = \sqrt{\frac{c_R}{c_F}}.$$

If $c_R/c_F \geq 1/4$, then $1/\theta^* = \infty$.

Table 3.2 presents optimum T_B^* and $1/\theta^*$ when $f(t) = [t^{\alpha-1}/(\alpha-1)!]e^{-t}$ ($\alpha = 2, 3, 4$) [8, p. 51] (Problem 3 in Sect. 3.4), i.e.,

$$F(t) = 1 - \sum_{j=0}^{\alpha-1} \frac{t^j}{j!}e^{-t},$$

$$M(t) = \sum_{k=1}^{\infty} \sum_{j=k\alpha}^{\infty} \frac{t^j}{j!}e^{-t}, \quad m(t) = \sum_{k=1}^{\infty} \frac{t^{k\alpha-1}}{(k\alpha-1)!}e^{-t}.$$

Both optimum T_B^* and $1/\theta^*$ increase with c_T/c_F or c_R/c_F, however, they do not have the monotone property for α. It is of interest that when $c_T = c_R$, $T_B^* > 1/\theta^*$ for small c_T/c_F, and conversely, $T_B^* < 1/\theta^*$ for large c_T/c_F. Naturally, it is unnecessary to make planned replacements, as the replacement costs c_T and c_R approach to the failure cost c_F.   □

Next, when the unit fails, it is not replaced and remains in a failed state for the time interval from a failure to its replacement [2, p. 120]. Then, the mean time from a failure to its replacement is

$$\int_0^{\infty} \left[\int_0^t (t-u)\,dF(u) \right] dG(t) = \int_0^{\infty} F(t)\overline{G}(t)\,dt.$$

Thus, the expected cost rate is

$$C_D(G) = \frac{c_D \int_0^\infty F(t)\overline{G}(t)\,dt + c_R}{\int_0^\infty \overline{G}(t)\,dt},$$ (3.22)

where c_D = downtime cost from failure to replacement. By similar arguments to periodic replacement (Problem 4 in Sect. 3.4), an optimum replacement is nonrandom, and the expected cost rate with replacement time T $(0 < T \le \infty)$ is

$$C_D(T) = \frac{c_D \int_0^T F(t)\,dt + c_T}{T},$$ (3.23)

where c_T = replacement cost at time T. Differentiating $C_D(T)$ with respect to T and setting it equal to zero,

$$\int_0^T t\,dF(t) = \frac{c_T}{c_D}.$$ (3.24)

Thus, if $\mu c_D > c_T$ then there exists a finite and unique T_D^* $(0 < T_D^* < \infty)$ which satisfies (3.24), and the resulting cost rate is

$$C_D(T_D^*) = c_D F(T_D^*).$$ (3.25)

In particular, when $F(t) = 1 - e^{-\lambda t}$ $(0 < \lambda < \infty)$, from (3.24), T_D^* is a unique solution of the equation

$$\frac{1}{\lambda}[1 - (1 + \lambda T)e^{-\lambda T}] = \frac{c_T}{c_D},$$ (3.26)

and the resulting cost rate is

$$C_D(T_D^*) = c_D(1 - e^{-\lambda T_D^*}).$$ (3.27)

Example 3.4 (Replacement for exponential random time) When $G(t) = 1 - e^{-\theta t}$ and $F(t) = 1 - e^{-\lambda t}$, the expected cost rate in (3.22) is

$$C_D(\theta) = \frac{c_D \lambda}{\theta + \lambda} + c_R \theta.$$ (3.28)

Thus, an optimum θ^* which minimizes $C_D(\theta)$ is given by

$$\left(\frac{\lambda}{\theta + \lambda}\right)^2 = \frac{\lambda c_R}{c_D},$$ (3.29)

and the expected cost rate is

$$C_D(\theta^*) = \frac{c_D\lambda}{\theta^* + \lambda}\frac{2\theta^* + \lambda}{\theta^* + \lambda}. \tag{3.30}$$

It has been shown that when $c_T = c_R$, periodic replacement is better than random one. We compute a random replacement cost $\widehat{c}_R$ when the expected costs of two replacement policies are the same one for $c_T > c_R$. From (3.27) and (3.30), we compute $1/\widehat{\theta}$ for c_T/c_D when

$$\frac{C_D^*(T_D^*)}{c_D} = 1 - e^{-\lambda T_D^*} = \frac{\lambda}{\widehat{\theta} + \lambda}\frac{2\widehat{\theta} + \lambda}{\widehat{\theta} + \lambda},$$

and compute

$$\frac{\widehat{c}_R}{c_D} = \frac{1}{\lambda}\left(\frac{\lambda}{\widehat{\theta} + \lambda}\right)^2.$$

It is noted that $T_D^* \to 0$ and $1/\theta^* \to 0$ as $c_T\ (> c_R) \to 0$. Thus, from (3.27) and (3.30),

$$\lim_{T\to 0}\frac{1 - e^{-\lambda T}}{[\lambda/(\lambda + 1/T)][(\lambda + 2/T)/(\lambda + 1/T)]} = \frac{1}{2}\lim_{T\to 0}\frac{1 - e^{-\lambda T}}{\lambda T} = \frac{1}{2}.$$

This shows that if $c_T \to 0$, then $C_D(T_D^*) \to C_D(\theta^*)/2$.

Table 3.3 presents optimum T_D^*, the resulting cost rates $C_D(T_D^*)/c_D$ and $C_D(\theta^*)/c_D$, and cost rate $\widehat{c}_R/c_T$ when both expected costs are the same. This indicates that the expected cost rate $C_D(T_D^*)$ approaches to the half of $C_D(\theta^*)$ as c_T becomes smaller.   □

Table 3.3 Values of T_D^*, $1/\widehat{\theta}$ and $\widehat{c}_R/c_D$ when $F(t) = 1 - e^{-t}$

c_T/c_D	T_D^*	$C_D(T_D^*)/c_D$	$1/\widehat{\theta}$	$\widehat{c}_R/c_D$	$\widehat{c}_R/c_T$	$C_D(\theta^*)/c_D$
0.001	0.045	0.044	0.023	0.001	1.000	0.062
0.002	0.065	0.063	0.033	0.001	0.500	0.087
0.005	0.103	0.098	0.053	0.003	0.600	0.136
0.010	0.148	0.138	0.077	0.005	0.500	0.190
0.020	0.215	0.193	0.113	0.010	0.500	0.263
0.050	0.356	0.299	0.195	0.027	0.540	0.397
0.100	0.532	0.412	0.305	0.055	0.550	0.532
0.200	0.823	0.561	0.509	0.114	0.570	0.694
0.500	1.675	0.813	1.311	0.322	0.644	0.914
1.000	∞	1.000	∞	1.000	1.000	1.000

Example 3.5 (Replacement for Uniform random time) Suppose that Y has a uniform distribution during $[0, T]$ given in Example 3.2. Then, the expected cost rate is, from (3.22),

$$C_U(T) = \frac{c_D \int_0^T (T - t)F(t)\,dt + c_R T}{T^2/2}. \tag{3.31}$$

Clearly,

$$\lim_{T \to 0} C_U(T) = \infty, \qquad \lim_{T \to \infty} C_U(T) = c_D.$$

Differentiating $C_U(T)$ with respect to T and setting it equal to zero,

$$\int_0^T \left(1 - \frac{2t}{T}\right) \overline{F}(t)\,dt = \frac{c_R}{c_D}, \tag{3.32}$$

whose left-hand side increases from 0 to μ. Thus, if $\mu c_D > c_R$, then there exists an optimum T_U^* $(0 < T_U^* < \infty)$ which satisfies (3.32), and the resulting cost rate is

$$C_U(T_U^*) = \frac{c_D \int_0^{T_U^*} F(t)\,dt + c_R}{T_U^*}, \tag{3.33}$$

which agrees with $C_D(T_U^*)$ in (3.23) and $T_U^* > T_D^*$ in (3.24) when $c_R = c_T$. In this case, the replacement with uniform random time is not better than standard one. □

In general, the above three kinds of random periodic policies appeared in this section are summarized as follows [2, p. 125]: The expected cost rate is

$$C(G) = \frac{c \int_0^\infty \overline{G}(t)\varphi(t)\,dt + c_R}{\int_0^\infty \overline{G}(t)\,dt}, \tag{3.34}$$

and when $G(t) = 1 - e^{-\theta t}$, an optimum θ^* satisfies

$$\int_0^\infty t\,e^{-\theta t}\,d\varphi(t) = \frac{c_R}{c}, \tag{3.35}$$

where $c = c_M, c_F, c_D$ and $\varphi(t) = h(t), m(t), F(t)$, respectively.

3.2 Random Periodic Replacement

The unit works for a job with a random time Y and undergoes minimal repair at
each failure. It has been shown in Sect. 3.1 that an optimum replacement policy is
nonrandom for an infinite time span. It might not be true in the case where units
work for a job with random working and processing times, because replacements
during its working times are impossible or impractical [1, p. 72], and replacement
costs after the completion of working times might be lower than those at planned
times. From such viewpoints, the replacement policy where the unit is replaced at
the first completion of working times over planned times was proposed [5]. We take
up four policies of standard replacement, replacement first, replacement last, and
replacement overtime, and compare them analytically and numerically.

We summarize the expected cost rates and their optimum replacement times for
four replacement policies. Next, we compare each policy with one another when
the working times are exponential and the replacement costs are the same. It is
shown similarly in Sect. 3.1 that standard replacement is the best among four ones,
as estimated previously. Furthermore, we determine either replacement first or last
is better than the other according to the ratio of replacement cost to minimal repair
cost. Finally, we discuss theoretically and numerically that if how much the cost for
random replacement is lower than that for periodic one, then random replacements
are the same as standard one.

3.2.1 Four Replacement Policies

We summarize the following four periodic and random replacement policies and
derive their optimum policies [9]:

3.2.1.1 Standard Replacement

A new unit begins to operate at time 0 and undergoes only minimal repair at each
failure. Suppose that the unit is replaced at periodic times kT $(k = 1, 2, \ldots)$ $(0 <
T \leq \infty)$, independently of its age, and any unit becomes as good as new after
replacement. The repair and replacement times are negligible. Then, the expected
cost rate is given in (3.2). If $\int_0^\infty t \, dh(t) > c_T/c_M$, then there exists a finite and unique
optimum T_S^* $(0 < T_S^* < \infty)$ which satisfies (3.4), and the resulting cost rate is given
in (3.5).

3.2.1.2 Replacement First

Suppose that the unit is replaced at time T or at time Y, whichever occurs first [2, p. 250]. Then, the mean time to replacements is

$$T\overline{G}(T) + \int_0^T t\,dG(t) = \int_0^T \overline{G}(t)\,dt,$$

and the expected number of failures until replacement is

$$H(T)\overline{G}(T) + \int_0^T H(t)\,dG(t) = \int_0^T \overline{G}(t)h(t)\,dt.$$

Thus, the expected cost rate is

$$C_F(T) = \frac{c_M \int_0^T \overline{G}(t)h(t)\,dt + c_R}{\int_0^T \overline{G}(t)\,dt}, \tag{3.36}$$

where c_R = replacement cost at time T or Y and c_M is given in (3.2). Differentiating $C_F(T)$ with respect to T and setting it equal to zero,

$$\int_0^T \overline{G}(t)[h(T) - h(t)]\,dt = \frac{c_R}{c_M},$$

or

$$\int_0^T \left[\int_0^t \overline{G}(u)\,du\right] dh(t) = \frac{c_R}{c_M}, \tag{3.37}$$

whose left-hand side increases strictly from 0 to $\int_0^\infty [\int_0^t \overline{G}(u)\,du]dh(t)$. Therefore, if $\int_0^\infty [\int_0^t \overline{G}(u)\,du]dh(t) > c_R/c_M$, then there exists an optimum T_F^* ($0 < T_F^* < \infty$) which satisfies (3.37), and the resulting cost rate is

$$C_F(T_F^*) = c_M h(T_F^*). \tag{3.38}$$

Note that if $h(\infty) = \infty$, then $\int_0^\infty [\int_0^t \overline{G}(u)\,du]dh(t) = \infty$ (Problem 5 in Sect. 3.4).

3.2.1.3 Replacement Last

Suppose that the unit is replaced at time T or at time Y, whichever occurs last [5]. Then, the mean time to replacement is

$$TG(T) + \int_T^\infty t \, dG(t) = T + \int_T^\infty \overline{G}(t) \, dt,$$

and the expected number of failures until replacement is

$$H(T)G(T) + \int_T^\infty H(t) \, dG(t) = H(T) + \int_T^\infty \overline{G}(t)h(t) \, dt.$$

Thus, the expected cost rate is

$$C_L(T) = \frac{c_M[H(T) + \int_T^\infty \overline{G}(t)h(t) \, dt] + c_R}{T + \int_T^\infty \overline{G}(t) \, dt}, \tag{3.39}$$

where c_M and c_R are given in (3.36). Differentiating $C_L(T)$ with respect to T and setting it equal to zero,

$$\int_0^T [h(T) - h(t)] \, dt - \int_T^\infty \overline{G}(t)[h(t) - h(T)] \, dt = \frac{c_R}{c_M}, \tag{3.40}$$

whose left-hand side increases strictly from $-\int_0^\infty \overline{G}(t)h(t) \, dt$ to $\int_0^\infty t \, dh(t)$. Therefore, if $\int_0^\infty t \, dh(t) > c_R/c_M$, then there exists a finite and unique T_L^* $(0 < T_L^* < \infty)$ which satisfies (3.40), and the resulting cost rate is

$$C_L(T_L^*) = c_M h(T_L^*). \tag{3.41}$$

3.2.1.4 Replacement Overtime

The unit works for a job with successive random times Y_j, each of which has an identical distribution $\Pr\{Y_j \le t\} \equiv G(t)$ with finite mean $1/\theta$ $(0 < \theta < \infty)$, as shown in Sect. 2.2.3. Suppose that the unit is replaced at the first completion of working times over time T. Then, the mean time to replacement is

$$\sum_{j=0}^\infty \int_0^T \left[\int_T^\infty u \, dG(u - t) \right] dG^{(j)}(t) \quad = T + \int_T^\infty \overline{G}(t) \, dt + \int_0^T \left[\int_T^\infty \overline{G}(u - t) \, du \right] dM(t),$$

and the expected number of failures until replacement is (Problem 6 in Sect. 3.4)

$$\sum_{j=0}^{\infty} \int_0^T \left[\int_T^{\infty} H(u) \, dG(u-t) \right] dG^{(j)}(t)$$

$$= H(T) + \int_T^{\infty} \overline{G}(t)h(t) \, dt + \int_0^T \left[\int_T^{\infty} \overline{G}(u-t)h(u) \, du \right] dM(t).$$

Therefore, the expected cost rate is

$$C_O(T) = \frac{c_M\{H(T) + \int_T^{\infty} \overline{G}(t)h(t) \, dt + \int_0^T [\int_T^{\infty} \overline{G}(u-t)h(u) \, du]dM(t)\} + c_R}{T + \int_T^{\infty} \overline{G}(t) \, dt + \int_0^T [\int_T^{\infty} \overline{G}(u-t) \, du]dM(t)},$$

(3.42)

where c_M and c_R are given in (3.36). Differentiating $C_O(T)$ with respect to T and setting it equal to zero,

$$\int_0^{\infty} \theta\overline{G}(t)h(T+t) \, dt \left\{ T + \int_T^{\infty} \overline{G}(t) \, dt + \int_0^T \left[\int_T^{\infty} \overline{G}(u-t) \, du \right] dM(t) \right\}$$

$$- \left\{ H(T) + \int_T^{\infty} \overline{G}(t)h(t) \, dt + \int_0^T \left[\int_T^{\infty} \overline{G}(u-t)h(u) \, du \right] dM(t) \right\} = \frac{c_R}{c_M},$$

i.e.,

$$\int_0^{\infty} \theta\overline{G}(t) \left(Th(T+t) - H(T) + \int_T^{\infty} \overline{G}(u)[h(T+t) - h(u)]du \right.$$

$$\left. + \int_0^T \left[\int_T^{\infty} \overline{G}(u-x)[h(T+t) - h(u)]du \right] dM(x) \right) dt = \frac{c_R}{c_M}, \quad (3.43)$$

whose left-hand increases strictly from 0 to $\int_0^{\infty} t \, dh(t)$ (Problem 7 in Sect. 3.4). Therefore, if $\int_0^{\infty} t \, dh(t) > c_R/c_M$, then there exists a finite and unique T_O^* ($0 < T_O^* < \infty$) which satisfies (3.43), and the resulting cost rate is

$$C_O(T_O^*) = c_M \int_0^{\infty} \theta\overline{G}(t)h(T_O^* + t) \, dt. \quad (3.44)$$

3.2.2 *Comparisons of Optimum Policies*

When $G(t) = 1 - e^{-\theta t}$, $c_T = c_R$ and the failure rate $h(t)$ increases strictly from 0 to ∞, we compare the above four policies. In this case, there exist every finite optimum T_S^*, T_F^*, T_L^* and T_O^*, which satisfy (3.4), (3.37), (3.40) and (3.43), respectively.

3.2.2.1 Comparisons of T_S^* and T_F^*, T_L^*, T_O^*

When $G(t) = 1 - e^{-\theta t}$ and $c_T = c_R$, from (3.37), an optimum T_F^* satisfies

$$\int_0^T e^{-\theta t}[h(T) - h(t)]\,dt = \frac{c_T}{c_M}, \tag{3.45}$$

and increases with θ from T_S^* to ∞, i.e., $T_F^* > T_S^*$. Thus, from (3.5) and (3.38), $C_S(T_S^*) < C_F(T_F^*)$. Therefore, standard replacement is better than replacement first.
Similarly, from (3.40), an optimum T_L^* satisfies

$$\int_0^T [h(T) - h(t)]\,dt - \int_T^\infty e^{-\theta t}[h(t) - h(T)]\,dt = \frac{c_T}{c_M}, \tag{3.46}$$

and decreases with θ from ∞ to T_S^*, i.e., $T_L^* > T_S^*$. Thus, from (3.5) and (3.41), $C_S(T_S^*) < C_L(T_L^*)$. Therefore, standard replacement is better than replacement last.
The expected cost rate of replacement overtime is, from (3.42),

$$C_O(T) = \frac{c_M[H(T) + \int_0^\infty e^{-\theta t}h(t+T)\,dt] + c_T}{T + 1/\theta}. \tag{3.47}$$

From (3.43), an optimum T_O^* satisfies

$$T\int_0^\infty \theta e^{-\theta t}h(t+T)\,dt - H(T) = \frac{c_T}{c_M}, \tag{3.48}$$

and increases with θ from 0 to T_S^* (Problem 8 in Sect. 3.4). The resulting cost rate is

$$C_O(T_O^*) = c_M \int_0^\infty \theta e^{-\theta t}h(t+T_O^*)\,dt = \frac{c_M H(T_O^*) + c_T}{T_O^*}. \tag{3.49}$$

It can be also seen that from (3.4) and (3.48), for $0 < T < \infty$,

$$\int_{0}^{\infty} \theta e^{-\theta t}[Th(t+T) - H(T)]\,dt > Th(T) - H(T),$$

which follows that $T_O^* < T_S^*$. In addition, because T_S^* minimizes the expected cost rate $C_S(T)$ in (3.2), $C_S(T_S^*) < C_O(T_O^*)$. Therefore, standard replacement is better than replacement overtime.

In three cases, when the replacement costs for four policies are the same, i.e., $c_T = c_R$, standard replacement is better than the other ones.

3.2.2.2 Comparisons of T_O^* and T_F^*, T_L^*

Compare T_O^* with T_F^*: From (3.45) and (3.48),

$$T\int_{0}^{\infty} \theta e^{-\theta t} h(t+T)\,dt - H(T) - \int_{0}^{T} e^{-\theta t}[h(T) - h(t)]\,dt$$

$$= T\int_{0}^{\infty} \theta e^{-\theta t}[h(t+T) - h(T)]\,dt + \int_{0}^{T}(1 - e^{-\theta t})[h(T) - h(t)]\,dt > 0,$$

and hence, $T_O^* < T_F^*$. Thus, we can compare $C_O(T_O^*)$ in (3.49) with $C_F(T_F^*)$ in (3.38) and determine which policy is better. For example, when $H(t) = \lambda t^2$, i.e., $h(t) = 2\lambda t$, from (3.49),

$$C_O(T_O^*) = 2c_M\lambda\left(T_O^* + \frac{1}{\theta}\right),$$

and from (3.38),

$$C_F(T_F^*) = 2c_M\lambda T_F^*.$$

Thus, if $T_O^* + 1/\theta < T_F^*$, then replacement overtime is better than replacement first, and *vice versa*.

Compare T_O^* with T_L^*: From (3.48) and (3.46),

$$T\int_{0}^{\infty} \theta e^{-\theta t} h(t+T)\,dt - H(T) - \int_{0}^{T}[h(T) - h(t)]\,dt + \int_{T}^{\infty} e^{-\theta t}[h(t) - h(T)]\,dt$$

$$= T\int_{0}^{\infty} \theta e^{-\theta t}[h(t+T) - h(T)]\,dt + \int_{T}^{\infty} e^{-\theta t}[h(t) - h(T)]\,dt > 0,$$

and hence, $T_O^* < T_L^*$. Thus, we can compare $C_O(T_O^*)$ in (3.49) with $C_L(T_L^*)$ in (3.41) and determine which policy is better. For example, when $h(t) = 2\lambda t$, from (3.41),

$$C_L(T_L^*) = 2c_M \lambda T_L^*.$$

Thus, if $T_O^* + 1/\theta < T_L^*$, then replacement overtime is better than replacement last and *vice versa* (Problem 9 in Sect. 3.4).

3.2.2.3 Comparison of T_F^* and T_L^*

From (3.45) and (3.46),

$$Q(T) \equiv \int_0^T [h(T) - h(t)]\, dt - \int_T^\infty e^{-\theta t}[h(t) - h(T)]dt - \int_0^T e^{-\theta t}[h(T) - h(t)]dt$$

$$= \int_0^T (1 - e^{-\theta t})[h(T) - h(t)]dt - \int_T^\infty e^{-\theta t}[h(t) - h(T)]dt, \qquad (3.50)$$

which increases strictly from $-\int_0^\infty e^{-\theta t}h(t)\, dt$ to ∞. Thus, there exists a finite and unique T_P ($0 < T_P < \infty$) which satisfies $Q(T) = 0$, and T_P decreases with θ from ∞ to 0 (Problem 10 in Sect. 3.4).

Denote that

$$L(T_P) \equiv \int_0^{T_P} e^{-\theta t}[h(T_P) - h(t)]dt, \qquad (3.51)$$

which decreases with θ to 0. Then, from (3.38) and (3.41), if $L(T_P) \geq c_R/c_M$ then $T_F^* \leq T_L^* \leq T_P$, and hence, replacement first is better than replacement last. Conversely, if $L(T_P) < c_R/c_M$ then $T_F^* > T_L^* > T_P$, and hence, replacement last is better than replacement first. In other words, the unit should be replaced earlier as the replacement cost and θ are smaller.

Example 3.6 (Replacement for Weibull failure time) Suppose that $c_T = c_R$, the failure time X has a Weibull distribution $F(t) = 1 - \exp(-t^2)$ and a random working time Y has an exponential distribution $G(t) = 1 - e^{-\theta t}$. Tables 3.4 and 3.5 present optimum T_F^*, T_L^*, and T_S^*, which satisfy (3.45), (3.46) and (3.4), respectively, and their cost rates for $1/\theta$ and c_T/c_M. Clearly, T_F^* and T_L^* become T_S^* when $1/\theta = \infty$ and $1/\theta = 0$, respectively. In addition, Table 3.4 gives T_P which satisfies $Q(T) = 0$ in (3.50) and $L(T_P)$ in (3.51).

Tables 3.4 and 3.5 indicate as follows:

Table 3.4 Optimum T_F^* and T_S^* and their cost rates

$\dfrac{c_T}{c_M}$	$1/\theta = 0.1$		$1/\theta = 0.2$		$1/\theta = 0.5$		$1/\theta = 1.0$		$1/\theta = \infty$	
	T_F^*	$\dfrac{C_F(T_F^*)}{c_M}$	T_F^*	$\dfrac{C_F(T_F^*)}{c_M}$	T_F^*	$\dfrac{C_F(T_F^*)}{c_M}$	T_F^*	$\dfrac{C_F(T_F^*)}{c_M}$	T_S^*	$\dfrac{C_S(T_S^*)}{c_M}$
0.01	0.120	0.240	0.109	0.218	0.103	0.206	0.102	0.204	0.100	0.200
0.02	0.184	0.368	0.160	0.320	0.148	0.296	0.145	0.290	0.141	0.282
0.05	0.347	0.694	0.274	0.548	0.242	0.484	0.233	0.466	0.224	0.448
0.10	0.600	1.200	0.426	0.852	0.353	0.706	0.334	0.668	0.316	0.632
0.20	1.100	2.200	0.694	1.388	0.525	1.050	0.483	0.966	0.447	0.894
0.50	2.600	5.200	1.450	2.900	0.921	1.842	0.801	1.602	0.707	1.414
1.00	5.100	10.200	2.700	5.400	1.474	2.948	1.198	2.396	1.000	2.000
2.00	10.100	20.200	5.200	10.400	2.497	4.994	1.841	3.682	1.414	2.828
5.00	25.100	50.200	12.700	25.400	5.500	11.000	3.469	6.938	2.236	4.472
T_P	0.130		0.260		0.650		1.230			
$L(T_P)$	0.011		0.046		0.286		1.045			

Table 3.5 Optimum T_L^* and its cost rate

$\dfrac{c_T}{c_M}$	$1/\theta = 0.1$		$1/\theta = 0.2$		$1/\theta = 0.5$		$1/\theta = 1.0$		$1/\widehat{\theta}$
	T_L^*	$\dfrac{C_L(T_L^*)}{c_M}$	T_L^*	$\dfrac{C_L(T_L^*)}{c_M}$	T_L^*	$\dfrac{C_L(T_L^*)}{c_M}$	T_L^*	$\dfrac{C_L(T_L^*)}{c_M}$	
0.01	0.125	0.250	0.199	0.398	0.458	0.916	0.905	1.810	0.093
0.02	0.156	0.312	0.217	0.434	0.466	0.932	0.909	1.818	0.132
0.05	0.228	0.456	0.267	0.534	0.488	0.976	0.920	1.840	0.209
0.10	0.318	0.636	0.339	0.678	0.525	1.050	0.939	1.878	0.296
0.20	0.447	0.894	0.456	0.912	0.593	1.186	0.976	1.952	0.418
0.50	0.707	1.414	0.709	1.418	0.778	1.556	1.084	2.168	0.661
1.00	1.000	2.000	1.000	2.000	1.031	2.062	1.253	2.506	0.935
2.00	1.414	2.828	1.414	2.828	1.424	2.848	1.556	3.112	1.322
5.00	2.236	4.472	2.236	4.472	2.237	4.474	2.281	4.562	2.090

(a) Both T_F^* and T_L^* increase with c_T/c_M. When c_T/c_M increases, the replacement time should be longer to lessen a high replacement cost, and replacement last is much better than replacement first, especially for small $1/\theta$. For example, when $1/\theta = 0.5$ and $c_T = 2c_M$, $T_L^* = 1.424$ is much less than $T_F^* = 2.497$.

(b) When $L(T_P) \geq c_T/c_M$, $T_F^* \leq T_L^*$ and replacement first is better than replacement last, and conversely, when $L(T_P) < c_T/c_M$, $T_F^* > T_L^*$ and replacement last is better than replacement first. For example, when $1/\theta = 0.5$, $L(T_P) = 0.286$, and hence, $T_F^* = 0.525 < T_L^* = 0.593$ for $c_T/c_M = 0.20$, and $T_L^* = 0.778 < T_F^* = 0.921$ for $c_T/c_M = 0.50$.

(c) Optimum T_F^* decreases to T_S^* with $1/\theta$ and T_L^* increases from T_S^* with $1/\theta$, because the unit is replaced at time $\min\{T_F^*, Y\}$ for replacement first, and at time $\max\{T_L^*, Y\}$ for replacement last. Furthermore, replacement first is better than replacement last as $1/\theta$ becomes larger. For example, when $c_T/c_M = 0.10$, if

$1/\theta \le 0.2$, then replacement last is better than replacement first, and if $1/\theta \ge 0.5$, then replacement first is better than replacement last. When $c_T/c_M = 0.50$ and $1/\theta = 1.0$, $T_F^* = 0.801 < 1/\theta = 1.0 < T_L^* = 1.084$, and optimum replacement times are equal nearly to $1/\theta$. Table 3.5 also presents $1/\theta$ such that both expected costs of replacement first and last are the same: From $Q(T_P) = 0$ in (3.50), θT_P is given by

$$\frac{(\theta T_P)^2}{2} - \theta T_P + 1 = 2e^{-\theta T_P}.$$

Using θT_P and $L(T_P)$ in (3.51), $\widehat{\theta}$ is computed for c_T/c_M by

$$2\left(\frac{1}{\widehat{\theta}}\right)^2 \left[\frac{(\widehat{\theta} T_P)^2}{2} - e^{-\widehat{\theta} T_P}\right] = \frac{c_T}{c_M}.$$

These values of $1/\widehat{\theta}$ increase with c_T/c_M. This indicates that when $c_T/c_M = 0.10$, if $1/\theta \le 0.296$, then replacement last is better than replacement first, and *vice versa*.   □

3.2.3 Comparisons of Policies with Different Replacement Costs

It has been shown that when $c_T = c_R$, standard replacement is better than the other ones. In general, the random replacement cost would be lower than the periodic one because the unit is replaced at random. We compute a random replacement cost c_R when the expected costs of the four replacement policies are the same.

3.2.3.1 Comparisons of T_S^* and T_F^*, T_L^*

First, compute T_S^* which satisfies (3.4). Next, compute c_{RF} and c_{RL} which satisfy, from (3.37) and (3.40), respectively,

$$\int_0^{T_S^*} \overline{G}(t)[h(T_S^*) - h(t)]dt = \frac{c_{RF}}{c_M},$$

$$\int_{T_S^*}^{\infty} \overline{G}(t)[h(t) - h(T_S^*)]dt = \frac{c_T - c_{RL}}{c_M}.$$

For example, when $h(t) = 2t$ and $G(t) = 1 - e^{-\theta t}$, from (3.4),

$$T_S^* = \sqrt{\frac{c_T}{c_M}},$$

and c_{RF} and c_{RL} are respective solutions of the equations:

$$\frac{2}{\theta^2}[\theta T_S^* - (1 - e^{-\theta T_S^*})] = \frac{c_{RF}}{c_M},$$

$$\frac{2}{\theta^2}\left[\frac{(\theta T_S^*)^2}{2} - e^{-\theta T_S^*}\right] = \frac{c_{RL}}{c_M}.$$

Clearly, c_{RF} is positive, however, c_{RL} might be negative. In this case, replacement last is never better than standard replacement.

3.2.3.2 Comparison of T_S^* and T_O^*

Compute T_O^* which satisfies, from (3.48),

$$T\int_0^\infty \theta e^{-\theta t} h(t+T)\,dt - H(T) = \frac{c_R}{c_M}.$$

Thus, from (3.5) and (3.49), if

$$h(T_S^*) > \int_0^\infty \theta e^{-\theta t} h(t + T_O^*)\,dt,$$

i.e.,

$$\frac{c_M H(T_S^*) + c_T}{T_S^*} > \frac{c_M H(T_O^*) + c_R}{T_O^*},$$

then replacement overtime is better than standard replacement. For example, when $h(t) = 2t$,

$$T_S^* = \sqrt{\frac{c_T}{c_M}}, \quad T_O^* = -\frac{1}{\theta} + \sqrt{\frac{1}{\theta^2} + \frac{c_R}{c_M}}.$$

Thus, if

$$T_S^* > T_O^* + \frac{1}{\theta} \quad \text{or} \quad \frac{c_T}{c_M} > \frac{1}{\theta^2} + \frac{c_R}{c_M},$$

Table 3.6 Optimum T_O^* and its cost rate

$\dfrac{c_T}{c_M}$	$1/\theta = 0.1$		$1/\theta = 0.2$		$1/\theta = 0.5$		$1/\theta = 1.0$	
	T_O^*	$\dfrac{C_O(T_O^*)}{c_M}$	T_O^*	$\dfrac{C_O(T_O^*)}{c_M}$	T_O^*	$\dfrac{C_O(T_O^*)}{c_M}$	T_O^*	$\dfrac{C_O(T_O^*)}{c_M}$
0.01	0.041	0.283	0.024	0.447	0.010	1.020	0.005	2.010
0.02	0.073	0.346	0.045	0.490	0.020	1.039	0.010	2.020
0.05	0.145	0.490	0.100	0.600	0.048	1.095	0.025	2.049
0.10	0.232	0.663	0.174	0.748	0.092	1.183	0.049	2.098
0.20	0.358	0.917	0.290	0.980	0.171	1.342	0.095	2.191
0.50	0.614	1.428	0.535	1.470	0.366	1.732	0.225	2.450
1.00	0.905	2.010	0.820	2.040	0.618	2.236	0.414	2.828
2.00	1.318	2.835	1.228	2.857	1.000	3.000	0.732	3.464
5.00	2.138	4.477	2.045	4.490	1.791	4.583	1.450	4.899

then replacement overtime is better than standard replacement. It can be clearly seen that when $c_R/c_M = c_T/c_M - 1/\theta^2$, both replacements are the same.

Example 3.7 (Replacement for Weibull failure time) We show the same numerical examples as those in Example 3.6 when $c_T = c_R$, $F(t) = 1 - \exp(-t^2)$ and $G(t) = 1 - e^{-\theta t}$. Table 3.6 presents optimum T_O^* and its cost rate $C_O(T_O^*)/c_M$ for $1/\theta$ and c_T/c_M. Optimum T_O^* increases with both c_T/c_M and θ. Compared to T_S^* in Table 3.4, $T_O^* + 1/\theta > T_S^*$, however, $T_O^* + 1/\theta$ approaches to T_S^* as c_T/c_M and θ increase. For example, when $c_T/c_M = 5.00$ and $1/\theta = 0.1$, $T_O^* + 1/\theta = 2.238 > T_S^* = 2.236$, however, when $c_T/c_M = 0.01$ and $1/\theta = 1.0$, $T_O^* + 1/\theta = 1.005 > T_S^* = 0.100$.

Compared to T_F^* in Table 3.4, replacement overtime becomes better than replacement first as both c_T/c_M and θ increase. On the other hand, compared to T_L^* in Table 3.5, $C_L(T_L^*) < C_O(T_O^*)$, however, $T_O^* + 1/\theta$ is a little greater than T_L^*.

When $c_T > c_R$, Table 3.7 presents c_{RF} and c_{RL} given in Sect. 3.2.3.1. Values of c_{RF} exist for all $1/\theta$ and c_T/c_M, however, the differences between c_{RF} and c_T become much smaller as c_T/c_M and θ are smaller. That is, if θ and c_T/c_M are small, standard replacement and replacement first become almost the same policy because the unit is replaced mainly at time T. Values of c_{RL} exist for large c_T/c_M and θ. In other words, replacement last cannot be rather than standard replacement when c_T/c_M and θ are small. Furthermore, if c_T/c_M and θ are large, c_{RL} becomes equal to c_T, i.e., both replacements are almost the same, because the unit is replaced mainly at time T. For example, when $1/\theta = 0.1$ and $c_T/c_M \geq 0.05$, c_{RL} and c_T are almost the same.   □

Table 3.7 Values of c_{RF}/c_M and c_{RL}/c_M when $C_S(T_S^*) = C_F(T_F^*)$ and $C_S(T_S^*) = C_L(T_L^*)$

$\dfrac{c_T}{c_M}$	$1/\theta = 0.1$		$1/\theta = 0.5$		$1/\theta = 1.0$	
	c_{RF}/c_M	c_{RL}/c_M	c_{RF}/c_M	c_{RL}/c_M	c_{RF}/c_M	c_{RL}/c_M
0.01	0.007	0.003	0.009	–	0.010	–
0.02	0.013	0.015	0.018	–	0.019	–
0.05	0.027	0.048	0.043	–	0.046	–
0.10	0.044	0.099	0.082	–	0.090	–
0.20	0.070	0.200	0.152	–	0.173	–
0.50	0.121	0.500	0.329	0.378	0.400	–
1.00	0.180	1.000	0.568	0.932	0.736	0.264
2.00	0.263	2.000	0.944	1.970	1.315	1.541
5.00	0.427	5.000	1.742	4.944	2.686	4.786

3.3 *N*th Working Time

Suppose that the unit is replaced at a planned number N ($N = 1, 2, \ldots$), i.e., at time Y_N in Sect. 2.4. Then, the probability that the unit works exactly j times in $[0, t]$ is $G^{(j)}(t) - G^{(j+1)}(t)$ [8, p. 50]. In addition, it is assumed that $G^{(j)}(t)$ has a density function $g^{(j)}(t)$, i.e., $g^{(j)}(t) \equiv dG^{(j)}(t)/dt$, $r_j(t) \equiv g^{(j)}(t)/[1-G^{(j)}(t)]$ ($j = 1, 2, \ldots$). Note that $r_j(t)dt$ represents the probability that the jth work of the unit finishes during $[t, t + dt]$, given that it operates at the jth number of working times at time t.

We take up the following three policies of replacement first, replacement last, and replacement overtime. When $N = 1$, these policies correspond to replacement policies in Sect. 3.2.

3.3.1 Replacement First

Suppose that the unit is replaced at time T ($0 < T \le \infty$) or at number N ($N = 1, 2, \ldots$), whichever occurs first. The probability that the unit is replaced at number N is $G^{(N)}(T)$, and the probability that it is replaced at time T is $1 - G^{(N)}(T)$. Thus, the mean time to replacement is

$$T[1 - G^{(N)}(T)] + \int_0^T t\, dG^{(N)}(t) = \int_0^T [1 - G^{(N)}(t)]dt,$$

and the total expected number of failures before replacement is

$$H(T)[1 - G^{(N)}(T)] + \int_0^T h(t)\, dG^{(N)}(t) = \int_0^T [1 - G^{(N)}(t)]h(t)\, dt.$$

Therefore, the expected cost rate is [5]

$$C_F(T, N) = \frac{c_M \int_0^T [1 - G^{(N)}(t)]h(t)\, dt + c_T + (c_R - c_T)G^{(N)}(T)}{\int_0^T [1 - G^{(N)}(t)]dt}, \qquad (3.52)$$

where c_M = minimal repair cost at each failure, c_T = replacement cost at time T, and c_R = replacement cost at number N. In general, c_R might depend on the replacement number N of working times. However, to simplify the model, c_R is assumed to be constant for any N in the average meaning.

When $N = \infty$, $C_F(T, \infty)$ agrees with $C_S(T)$ in (3.2), and when $N = 1$, $C_F(T, 1)$ agrees with $C_F(T)$ in (3.36) for $c_R = c_T$. Furthermore, when the unit is replaced only at number N,

$$C_F(N) \equiv \lim_{T \to \infty} C_F(T, N) = \frac{c_M \int_0^\infty [1 - G^{(N)}(t)]h(t)\, dt + c_R}{N/\theta} \qquad (N = 1, 2, \ldots).$$
$$(3.53)$$

When $N = 1$, $C_F(N)$ agrees with $C_R(G)$ in (3.1). First, we derive an optimum number N_F^* which minimizes $C_F(N)$. From the inequality $C_F(N + 1) - C_F(N) \geq 0$,

$$\frac{N}{\theta} H_N - \int_0^\infty [1 - G^{(N)}(t)]h(t)\, dt \geq \frac{c_R}{c_M}, \qquad (3.54)$$

where

$$H_N(T) \equiv \frac{\int_0^T [G^{(N)}(t) - G^{(N+1)}(t)]h(t)\, dt}{\int_0^T [G^{(N)}(t) - G^{(N+1)}(t)]dt} \leq h(T),$$

and

$$H_N \equiv \lim_{T \to \infty} H_N(T) = \int_0^\infty \theta[G^{(N)}(t) - G^{(N+1)}(t)]h(t)\, dt.$$

Thus, if H_N increases strictly, then the left-hand side of (3.54) increases strictly (Problem 11 in Sect. 3.4). So that, if there exists some N such that (3.54) holds, an optimum number N_F^* is given by a finite and unique minimum which satisfies (3.54). When $G(t) = 1 - e^{-\theta t}$,

$$H_N = \int_0^\infty \frac{\theta(\theta t)^N}{N!} e^{-\theta t} h(t)\, dt \quad (N = 0, 1, 2, \ldots)$$

increases strictly with N to $h(\infty)$ [7, p. 160] from (12) of Appendix A.2. In this case, the left-hand side of (3.54) is

$$\int_0^\infty [1 - G^{(N)}(t)][H_N - h(t)]dt,$$

and increases strictly with N to

$$\frac{1}{\theta} \sum_{j=0}^\infty [h(\infty) - H_j]. \tag{3.55}$$

Therefore, if (3.55) is greater than c_R/c_M, then there exists a finite and unique minimum N_F^* ($1 \le N_F^* < \infty$) which satisfies (3.54).

Second, we discuss both optimum T_F^* and N_F^* which minimize $C_F(T, N)$ in (3.52). Differentiating $C_F(T, N)$ with respect to T and setting it equal to zero,

$$c_M \int_0^T [1 - G^{(N)}(t)][h(T) - h(t)]dt$$

$$- (c_T - c_R) \int_0^T [1 - G^{(N)}(t)][r_N(T) - r_N(t)]dt = c_T. \tag{3.56}$$

From the inequality $C_F(T, N+1) - C_F(T, N) \ge 0$,

$$c_M \int_0^T [1 - G^{(N)}(t)][H_N(T) - h(t)]dt$$

$$+ (c_T - c_R) \int_0^T [1 - G^{(N)}(t)] \left\{ \frac{G^{(N)}(T) - G^{(N+1)}(T)}{\int_0^T [G^{(N)}(t) - G^{(N+1)}(t)]dt} + r_N(t) \right\} dt \ge c_T.$$

$$\tag{3.57}$$

Substituting (3.56) for (3.57),

$$c_M[H_N(T) - h(T)] + (c_T - c_R)\left\{\frac{G^{(N)}(T) - G^{(N+1)}(T)}{\int_0^T [G^{(N)}(t) - G^{(N+1)}(t)]dt} + r_N(T)\right\} \geq 0.$$

(3.58)

Thus, when $c_T \leq c_R$, there dose not exist any finite optimum N_F^* for $T > 0$, i.e., $N_F^* = \infty$.

Next, we derive both optimum T_F^* and N_F^* when $h(\infty) = \infty$, $G(t) = 1 - e^{-\theta t}$ and $c_T > c_R$. Then, it is easily proved from (1) of Appendix A.1 that

$$r_N(T) = \frac{\theta(\theta T)^{N-1}/(N-1)!}{\sum_{j=0}^{N-1}[(\theta T)^j/j!]}$$

(3.59)

decreases strictly with N from θ to 0 and increases strictly with T from 0 to θ for $N \geq 2$, and

$$\frac{G^{(N)}(T) - G^{(N+1)}(T)}{\int_0^T [G^{(N)}(t) - G^{(N+1)}(t)]dt} = \frac{\theta(\theta T)^N/N!}{\sum_{j=N+1}^{\infty}[(\theta T)^j/j!]}$$

(3.60)

increases with N to ∞ and decreases with T from ∞ to 0. Thus, because $\lim_{N\to\infty} H_N(T) = h(T)$ from (12) of Appendix A.2, and from (3.59) and (3.60), there exists a finite N_F^* ($1 \leq N_F^* < \infty$) which satisfies (3.58) for $T > 0$. Furthermore, because $h(\infty) = \infty$, there exists a finite T_F^* ($0 < T_F^* < \infty$) which satisfies (3.56) for $N \geq 1$.

Example 3.8 (*Replacement for Weibull failure time*) Table 3.8 presents optimum N_F^* which minimizes $C_F(N)$ in (3.53), and (T_F^*, N_F^*) which minimize $C_F(T, N)$ in (3.52) when $G(t) = 1 - e^{-t}$, $h(t) = t/10$ and $c_M = 5$. This indicates that a finite N_F^* exists for $c_R < c_T$. For (T_F^*, N_F^*), N_F^* increases with c_R and decreases with c_T, however,

Table 3.8 Optimum N_F^* and (T_F^*, N_F^*) when $G(t) = 1 - e^{-t}$, $h(t) = t/10$ and $c_M = 5$

c_T	$c_R = 10$			$c_R = 15$		
	N_F^*	$(T_F^*$	$N_F^*)$	N_F^*	$(T_F^*$	$N_F^*)$
10	6	(6.325	∞)	8	(6.325	∞)
11	6	(7.112	8)	8	(6.633	∞)
12	6	(8.373	7)	8	(6.928	∞)
13	6	(9.607	7)	8	(7.211	∞)
14	6	(11.792	6)	8	(7.483	∞)
15	6	(13.450	6)	8	(7.746	∞)
16	6	(15.194	6)	8	(8.248	11)
17	6	(17.000	6)	8	(9.352	9)
18	6	(18.850	6)	8	(10.871	8)
19	6	(20.732	6)	8	(12.179	8)
20	6	(22.637	6)	8	(13.638	8)

T_F^* decreases with c_R and increases with c_T. It is of interest that when c_T is much higher than c_R, both N_F^* become equal and are less than T_F^*, i.e., we might replace the unit only at number N_F^* because the replacement cost at number N is much lower than that at time T. □

3.3.2 Replacement Last

Suppose that the unit is replaced a time T $(0 \le T \le \infty)$ or at number N $(N = 0, 1, 2, \ldots)$, whichever occurs last. The probability that the unit is replaced at number N is $1 - G^{(N)}(T)$, and the probability that the unit is replaced at time T is $G^{(N)}(T)$. Thus, the mean time to replacement is

$$TG^{(N)}(T) + \int_T^\infty t \, dG^{(N)}(t) = T + \int_T^\infty [1 - G^{(N)}(t)]dt,$$

and the expected number of failures before replacement is

$$H(T)G^{(N)}(T) + \int_T^\infty H(t) \, dG^{(N)}(t) = H(T) + \int_T^\infty [1 - G^{(N)}(t)]h(t) \, dt.$$

By a similar method of obtaining (3.52), the expected cost rate is [5]

$$C_\mathrm{L}(T, N) = \frac{c_\mathrm{M}\{H(T) + \int_T^\infty [1 - G^{(N)}(t)]h(t) \, dt\} + c_\mathrm{R} + (c_T - c_\mathrm{R})G^{(N)}(T)}{T + \int_T^\infty [1 - G^{(N)}(t)]dt}.$$

$$(3.61)$$

In particular, when $N = 0$, $C_\mathrm{L}(T, 0) = C_\mathrm{S}(T)$ in (3.2), and when $N = 1$, $C_\mathrm{L}(T, 1) = C_\mathrm{L}(T)$ in (3.39) for $c_T = c_\mathrm{R}$. Furthermore, when $T = 0$, $C_\mathrm{L}(0, N) = C_\mathrm{F}(N)$ in (3.53).

We find both optimum T_L^* and N_L^* which minimize $C_\mathrm{L}(T, N)$ in (3.61). Differentiating $C_\mathrm{L}(T, N)$ with respect to T and setting it equal to zero,

$$c_\mathrm{M}\left\{ Th(T) - H(T) - \int_T^\infty [1 - G^{(N)}(t)][h(t) - h(T)] \, dt \right\}$$

$$+ (c_T - c_\mathrm{R})\left(\tilde{r}_N(T)\left\{ T + \int_T^\infty [1 - G^{(N)}(t)]dt \right\} - G^{(N)}(T) \right) = c_\mathrm{R}, \quad (3.62)$$

where $\tilde{r}_N(t) \equiv g^{(N)}(t)/G^{(N)}(t)$. From the inequality $C_\mathrm{L}(T, N+1) - C_\mathrm{L}(T, N) \ge 0$,

$$
c_M \left(\tilde{H}_N(T) \left\{ T + \int_T^\infty [1 - G^{(N)}(t)] dt \right\} - H(T) - \int_T^\infty [1 - G^{(N)}(t)] h(t) \, dt \right)
$$

$$
- (c_T - c_R) \left(\frac{G^{(N)}(T) - G^{(N+1)}(T)}{\int_T^\infty [G^{(N)}(t) - G^{(N+1)}(t)] dt} \left\{ T + \int_T^\infty [1 - G^{(N)}(t)] dt \right\} \right.
$$

$$
\left. + G^{(N)}(T) \right) \geq c_R, \tag{3.63}
$$

where

$$
\tilde{H}_N(T) \equiv \frac{\int_T^\infty [G^{(N)}(t) - G^{(N+1)}(t)] h(t) \, dt}{\int_T^\infty [G^{(N)}(t) - G^{(N+1)}(t)] dt} \geq h(T).
$$

Substituting (3.62) for (3.63),

$$
c_M \left[\tilde{H}_N(T) - h(T) \right] - (c_T - c_R) \left\{ \frac{G^{(N)}(T) - G^{(N+1)}(T)}{\int_T^\infty [G^{(N)}(t) - G^{(N+1)}(t)] dt} + \tilde{r}_N(T) \right\} \geq 0. \tag{3.64}
$$

Because $\tilde{H}_N(T) \geq h(T)$, there does not exist any N_L^* for $T > 0$ and $c_T \leq c_R$, i.e., $N_L^* = 0$. Furthermore, when $c_T > c_R$, $N = 0$ and $G(t) = 1 - e^{-\theta t}$, the left-hand side of (3.64) is simplified as

$$
c_M \int_0^\infty \theta e^{-\theta t} [h(t + T) - h(T)] dt - (c_T - c_R) \theta.
$$

Thus, if $\int_0^\infty e^{-\theta t} [h(t + T) - h(T)] dt < (c_T - c_R)/c_M$, then a positive N_L^* ($N_L^* \geq 1$) exists.

Example 3.9 (Replacement for Weibull failure time) Table 3.9 presents optimum (T_L^*, N_L^*) when $G(t) = 1 - e^{-t}$, $h(t) = t/10$ and $c_M = 5$. This indicates that both T_L^* and N_L^* increase with c_T, however, T_L^* increases with c_R for small c_T and decreases for large c_T, and N_L^* decreases with c_R. It is of interest that when c_T is large, T_L^* is almost the same as $N_L^* + 1$, i.e., $T_L^* \approx N_L^* + 1/\theta$. Compared to Tables 3.8 and 3.9, when $N_F^* = \infty$, $N_L^* = 0$, and $T_F^* = T_L^*$.   □

Table 3.9 Optimum (T_L^*, N_L^*) when $G(t) = 1 - e^{-t}$, $h(t) = t/10$ and $c_M = 5$

c_T	$c_R = 10$		$c_R = 15$	
	T_L^*	N_L^*	T_L^*	N_L^*
10	6.325	0	6.325	0
11	6.568	3	6.633	0
12	5.159	6	6.928	0
13	6.470	6	7.211	0
14	7.358	6	7.483	0
15	7.524	6	7.483	0
16	8.122	7	7.977	3
17	8.286	7	7.688	6
18	8.465	7	8.012	7
19	9.139	8	7.860	7
20	9.291	8	8.759	7

3.3.3 Replacement Overtime

Suppose that the unit is replaced at the Nth ($N = 1, 2, \ldots$) number of working times or at the first completion of working times over time T ($0 \le T \le \infty$), whichever occurs first. Then, the probability that the unit is replaced at number N before time T is $G^{(N)}(T)$, and the probability that it is replaced at the first completion of working times over time T is $1 - G^{(N)}(T)$. The mean time to replacement is

$$\int_0^T t\,dG^{(N)}(t) + \sum_{j=0}^{N-1} \int_0^T \left[\int_T^\infty u\,dG(u - t) \right] dG^{(j)}(t)$$

$$= \int_0^T [1 - G^{(N)}(t)]dt + \sum_{j=0}^{N-1} \int_0^T \left[\int_T^\infty \overline{G}(u - t)\,du \right] dG^{(j)}(t),$$

and the expected number of failures before replacement is

$$\int_0^T H(t)\,dG^{(N)}(t) + \sum_{j=0}^{N-1} \int_0^T \left[\int_T^\infty H(u)\,dG(u - t) \right] dG^{(j)}(t)$$

$$= \int_0^T [1 - G^{(N)}(t)]h(t)\,dt + \sum_{j=0}^{N-1} \int_0^T \left[\int_T^\infty \overline{G}(u - t)h(u)\,du \right] dG^{(j)}(t).$$

Therefore, the expected cost rate is

$$C_{OF}(T,N) = \frac{c_M\{\sum_{j=0}^{N-1}\int_0^T[\int_T^\infty \overline{G}(u-t)h(u)\,du]dG^{(j)}(t)}{+\int_0^T[1-G^{(N)}(t)]h(t)\,dt\} + c_T + (c_R - c_T)G^{(N)}(T)}{\int_0^T[1-G^{(N)}(t)]dt + \sum_{j=0}^{N-1}\int_0^T[\int_T^\infty \overline{G}(u-t)\,du]dG^{(j)}(t)}. \quad (3.65)$$

When $N = \infty$, $C_O(T,\infty) = C_O(T)$ in (3.42). Furthermore, when $T = \infty$, $C_O(\infty, N) = C_F(N)$ in (3.53) and $C_O(\infty, 1) = C_R(G)$ in (3.1) (Problem 12 in Sect. 3.4).

Next, suppose that the unit is replaced at number N or at the first completion of working times over time T, whichever occurs last. Then, the probability that the unit is replaced at number N after time T is $1 - G^{(N)}(T)$, and the probability that it is replaced at the first completion of working times over time T is $G^{(N)}(T)$. The mean time to replacement is

$$\int_T^\infty t\,dG^{(N)}(t) + \sum_{j=N}^\infty \int_0^T\left[\int_T^\infty u\,dG(u-t)\right]dG^{(j)}(t)$$

$$= T + \int_T^\infty[1-G^{(N)}(t)]dt + \sum_{j=N}^\infty \int_0^T\left[\int_T^\infty \overline{G}(u-t)\,du\right]dG^{(j)}(t),$$

and the expected number of failures before replacement is

$$H(T) + \int_T^\infty[1-G^{(N)}(t)]h(t)\,dt + \sum_{j=N}^\infty \int_0^T\left[\int_T^\infty \overline{G}(u-t)h(u)\,du\right]dG^{(j)}(t).$$

Therefore, the expected cost rate is (Problem 13 in Sect. 3.4)

$$C_{OL}(T,N) = \frac{c_M\{\sum_{j=N}^\infty \int_0^T[\int_T^\infty \overline{G}(u-t)h(u)\,du]dG^{(j)}(t)}{+H(T) + \int_T^\infty[1-G^{(N)}(t)]h(t)\,dt\} + c_R + (c_T - c_R)G^{(N)}(T)}{T + \int_T^\infty[1-G^{(N)}(t)]dt + \sum_{j=N}^\infty \int_0^T[\int_T^\infty \overline{G}(u-t)\,du]dG^{(j)}(t)}. \quad (3.66)$$

3.3.4 Replacement with Constant Time

Suppose that when $c_T = c_R$ and $F(t) = 1 - e^{-\theta t}$, i.e., $G^{(N)}(t) = \sum_{j=N}^\infty[(\theta t)^j/j!]e^{-\theta t}$ ($N = 0, 1, 2, \ldots$), a planned replacement time T ($0 < T < \infty$) is fixed. It would be estimated from the above discussions that if $T \leq T_S^*$, then the unit should be replaced only at time T. From (3.52), we find an optimum N_F^* of replacement first for a given T, which minimizes the expected cost rate

$$C_F(N; T) = \frac{c_M \int_0^T [1 - G^{(N)}(t)] h(t)\, dt + c_T}{\int_0^T [1 - G^{(N)}(t)]\, dt} \quad (N = 1, 2, \ldots). \tag{3.67}$$

From the inequality $C_F(N + 1; T) - C_F(N; T) \geq 0$, i.e., setting $c_T = c_R$ in (3.57),

$$\int_0^T [1 - G^{(N)}(t)][H_N(T; \theta) - h(t)]\, dt \geq \frac{c_T}{c_M}, \tag{3.68}$$

where

$$H_N(T; \theta) \equiv \frac{\int_0^T (\theta t)^N e^{-\theta t} h(t)\, dt}{\int_0^T (\theta t)^N e^{-\theta t}\, dt} \leq h(T).$$

From (3.4), (3.68) is rewritten as

$$\int_0^T [1 - G^{(N)}(t)][H_N(T; \theta) - h(t)]\, dt \geq \int_0^{T_S^*} [h(T_S^*) - h(t)]\, dt. \tag{3.69}$$

Recalling that $H_N(T; \theta)$ increases strictly with N to $h(T)$ from **(12)** of Appendix A.2, the left-hand of (3.69) also increases strictly with N to $\int_0^T [h(T) - h(t)]\, dt$. Thus, if $T \leq T_S^*$ then there does not exist any finite N which satisfies (3.68), i.e., $N_F^* = \infty$. Conversely, if $T > T_S^*$ then there exists a finite and unique minimum N_F^* which satisfies (3.68).

Next, from (3.61), we find an optimum N_L^* of replacement last for a given T which minimizes the expected cost rate

$$C_L(N; T) = \frac{c_M \{H(T) + \int_T^\infty [1 - G^{(N)}(t)] h(t)\, dt\} + c_T}{T + \int_T^\infty [1 - G^{(N)}(t)]\, dt} \quad (N = 0, 1, 2, \ldots). \tag{3.70}$$

From the inequality $C_L(N + 1; T) - C_L(N; T) \geq 0$, i.e., setting $c_T = c_R$ in (3.63),

$$\int_T^\infty [1 - G^{(N)}(t)][\tilde{H}_N(T; \theta) - h(t)]\, dt + \int_0^T [\tilde{H}_N(T; \theta) - h(t)]\, dt$$

$$\geq \int_0^{T_S^*} [h(T_S^*) - h(t)]\, dt \quad (N = 0, 1, 2, \ldots), \tag{3.71}$$

where

$$\tilde{H}_N(T;\theta) \equiv \frac{\int_T^\infty (\theta t)^N e^{-\theta t} h(t)\, dt}{\int_T^\infty (\theta t)^N e^{-\theta t}\, dt} \ge h(T).s$$

Recalling that $\tilde{H}_N(T;\theta)$ increases strictly with N to $h(\infty)$ and $\tilde{H}_N(T;\theta) \ge h(T)$ for $N \ge 0$ from (15) of Appendix A.2, the left-hand side also increases strictly with N from $\int_0^T [\tilde{H}_0(T;\theta) - h(t)]dt$. Thus, if $T \ge T_S^*$ then there does not exist any positive N which satisfies (3.71), i.e., $N_L^* = 0$. Conversely, if $T < T_S^*$ and $h(\infty) = \infty$, then there exists a finite and unique minimum N_L^* which satisfies (3.71) (Problem 14 in Sect. 3.4).

3.4 Problems

1. Show that the three equations in (3.4) are equivalent.
2. Prove that an optimum policy that minimizes $C_B(G)$ in (3.15) is nonrandom.
3. Compute T_B^* numerically when $f(t) = [t^{\alpha-1}/(\alpha-1)!]e^{-t}$.
4. Prove that an optimum policy that minimizes $C_D(G)$ in (3.22) is nonrandom.
5. Prove that if $h(\infty) = \infty$, then

$$\lim_{T\to\infty} \int_0^T \overline{G}(t)[h(T) - h(t)]dt = \infty.$$

6. Derive $C_O(T)$ in (3.42).
7. Prove that the left-hand side of (3.43) increases strictly from 0 to $\int_0^\infty t\, dh(t)$.
8. Prove that T_L^* decreases with θ from ∞ to T_S^*, and T_O^* increases with θ from 0 to T_S^*.
9. When $H(t) = \lambda t^3$, compare replacement overtime and replacement first and last.
10. Prove that a solution T_P of (3.50) decreases with θ from ∞ to 0.
11. Prove that if H_N increases strictly, then the left-hand side of (3.54) increases strictly, and when $G(t) = 1 - e^{-\theta t}$, H_N increases strictly to $h(\infty)$.
*12. When the unit is replaced at time T after the Nth working time and before the Nth time, it is replaced at the first completion of working time, obtain the expected cost rate, and derive an optimum policy.
*13. Discuss optimum policies which minimizes $C_{OF}(T,N)$ in (3.65) and $C_{OL}(T,N)$ in (3.66).
14. Make a numerical example of optimum N_F^* in (3.69) and N_L^* in (3.71) for a given T and a Weibull failure time.

References

1. Barlow RE, Proschan F (1965) Mathematical theory of reliability. Wiley, New York
2. Nakagawa T (2005) Maintenance theory of reliability. Springer, London
3. Tadj L, Ouali MS, Yacout S, Ait-Kadi S (eds) (2011) Replacement models with minimal repair. Springer, London
4. Chen M, Mizutani S, Nakagawa T (2010) Random and age replacement policies. Int J Reliab Qual Saf Eng 17:27–39
5. Chen M, Nakamura S, Nakagawa T (2010) Replacement and preventive maintenance models with random working times, IEICE Trans Fundam, E93-A: 500–507
6. Pinedo M (2002) Scheduling theory. Algorithms and Systems, Prentice Hall
7. Nakagawa T (2008) Advanced reliability models and maintenance policies. Springer, London
8. Nakagawa T (2011) Stochastic processes with applications to reliability theory. Springer, London
9. Zhao X, Nakagawa T (2014) Comparisons of periodic and random replacement policies. To appear in Frenkel I, et al. (eds) Applied Reliability Engineering and Risk Analysis, Probabilistic Models and Statistical Inference. Wiley, New York, pp 193–204

Chapter 4
Random Inspection Policies

Most units in standby and in storage have to be checked at suitable times to detect their failures, which is called an *inspection policy*. Optimum policies that minimize the total expected cost until failure detection were derived [1, p. 107]. Asymptotic inspection schedules and policies for standby and storage units were discussed extensively [2, p. 201], and their applications to gas pipelines and plants were shown [3, p. 423]. The delay time models of inspection were applied to plant maintenance [4, 5]. The periodic and sequential inspection policies for a finite time interval were summarized [6], [7, p. 64].

Some systems in offices and industries successively execute jobs and computer processes. For such systems, it would be impossible or impractical to maintain them in a strict periodic fashion, as shown in Chaps. 2 and 3. In this chapter, we consider the same operating system, which executes a job with random working times Y_j ($j = 1, 2, \ldots$), and $S_j \equiv \sum_{i=1}^{j} Y_i$, $S_0 \equiv 0$ given in Fig. 2.2. It is assumed that Y_j ($j = 1, 2, \ldots$) is independent, and has an identical distribution $G(t) \equiv \Pr\{Y_j \leq t\}$ with finite mean $1/\theta$ ($0 < \theta < \infty$). Then, the probability that the system works exactly j times in $[0, t]$ is $G^{(j)}(t) - G^{(j+1)}(t)$, where $G^{(j)}(t)$ ($j = 1, 2, \ldots$) denotes the j-fold Stieltjes convolution of $G(t)$ with itself and $G^{(0)}(t) \equiv 1$ for $t \geq 0$. In addition, $M(t) \equiv \sum_{j=1}^{\infty} G^{(j)}(t)$ represents the expected number of works in $[0, t]$.

Suppose that the unit deteriorates with its age, i.e., the total working time, irrespective of the number of works, and fails according to a general distribution $F(t)$ and its density function $f(t) \equiv dF(t)/dt$ with finite mean $\mu \equiv \int_0^{\infty} \overline{F}(t)dt$ ($0 < \mu < \infty$), where $\overline{\Phi}(t) \equiv 1 - \Phi(t)$ for any function $\Phi(t)$.

We apply the inspection policy to the unit with a random working time Y_j: It is assumed in Sect. 4.1 that the unit is checked at successive working times S_j and also at periodic times kT ($k = 1, 2, \ldots$) [8, 9]. The total expected costs until failure detection are obtained, and the optimum policies which minimize them for periodic and random inspections are derived, respectively. In addition, we compare periodic and random inspection policies when the failure time is exponential. It is shown that periodic inspection is better than random one when both costs of periodic and random inspections are the same. However, if the random inspection cost is the half

© Springer-Verlag London 2014
T. Nakagawa, *Random Maintenance Policies*,
Springer Series in Reliability Engineering, DOI 10.1007/978-1-4471-6575-0_4

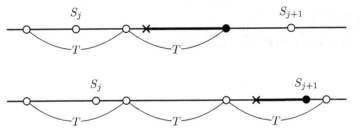

○ Check at periodic and random times   ✕ Failure time   ● Detection of failure

Fig. 4.1 Process of random and periodic inspections

of periodic one, both expected costs are almost the same. Furthermore, when the unit is checked at successive times T_k $(k = 1, 2, \ldots)$, optimum checking times are computed numerically.

It is assumed in Sect. 4.2 that the unit is checked at every completion of Nth $(N = 1, 2, \ldots)$ working time [8]. An optimum number N^* which minimizes the total expected cost is derived analytically. It is assumed in Sect. 4.3 that failure and working times are exponential. In similar ways of taking up maintenance policies in Chaps. 2 and 3, we propose three modified inspection policies where the unit is checked at a planned time T or at a working time Y, whichever occurs first or last, and at the first completion of working times, which are called *inspection first*, *inspection last*, and *inspection overtime*. These policies include the standard periodic and random inspection ones. We obtain the total expected costs of each policy and derive analytically optimum policies which minimize them. In addition, we compare inspection first and last. It is shown that either of them is better than the other according to the ratio of checking cost to downtime cost from a failure to its detection. Furthermore, we also compare periodic inspection and inspection overtime. Finally, we take up a random inspection policy for a finite interval in Sect. 4.4, and also, inspection policies for a random finite interval in Sect. 8.1.1.

4.1 Periodic and Random Inspections

Suppose that the unit is checked at successive working times S_j $(j = 1, 2, \ldots)$ and also at periodic times kT $(k = 1, 2, \ldots)$ for a specified T $(0 < T \leq \infty)$ in Fig. 4.1. The failure is certainly detected by either random or periodic checking times, whichever occurs first, and the process ends.

The probability that the failure is detected by periodic check is

$$\sum_{k=0}^{\infty} \int_{kT}^{(k+1)T} \left\{ \sum_{j=0}^{\infty} \int_0^t \overline{G}[(k+1)T - x] \, dG^{(j)}(x) \right\} dF(t), \qquad (4.1)$$

and the probability that it is detected by random check is

$$\sum_{k=0}^{\infty} \int_{kT}^{(k+1)T} \left(\sum_{j=0}^{\infty} \int_{0}^{t} \{G[(k+1)T - x] - G(t - x)\} \, dG^{(j)}(x) \right) dF(t), \qquad (4.2)$$

where $(4.1)+(4.2)=1$.

Let c_T be the cost for periodic check, c_R be the cost for random check, and c_D be the downtime cost per unit of time for the time elapsed between a failure and its detection at the next check. Then, the total expected cost until failure detection is [2, p. 254]

$$C(T) = \sum_{k=0}^{\infty} \int_{kT}^{(k+1)T} \left(\sum_{j=0}^{\infty} \{(k+1)c_T + jc_R + c_D[(k+1)T - t]\} \right.$$

$$\times \int_{0}^{t} \overline{G}[(k+1)T - x] \, dG^{(j)}(x) \right) dF(t)$$

$$+ \sum_{k=0}^{\infty} \int_{kT}^{(k+1)T} \left(\sum_{j=0}^{\infty} \int_{0}^{t} \left\{ \int_{t-x}^{(k+1)T-x} [kc_T + (j+1)c_R \right.\right.$$

$$+ c_D(x + y - t)] \, dG(y) \Big\} \, dG^{(j)}(x) \right) dF(t)$$

$$= c_T \sum_{k=0}^{\infty} \overline{F}(kT) + c_R \int_{0}^{\infty} M(t) \, dF(t) - (c_T - c_R) \sum_{k=0}^{\infty} \int_{kT}^{(k+1)T} \Bigg[G[(k+1)T]$$

$$- G(t) + \left(\int_{0}^{t} \{G[(k+1)T - x] - G(t - x)\} \, dM(x) \right) \Bigg] dF(t)$$

$$+ c_D \sum_{k=0}^{\infty} \int_{kT}^{(k+1)T} \left(\int_{t}^{(k+1)T} \overline{G}(y) \, dy \right.$$

$$+ \left. \int_{0}^{t} \left[\int_{t-x}^{(k+1)T-x} \overline{G}(y) \, dy \right] dM(x) \Big\} \right) dF(t), \qquad (4.3)$$

where $M(t) \equiv \sum_{j=1}^{\infty} G^{(j)}(t)$ represents the expected number of random checks during $(0, t]$. In particular, when $T = \infty$, i.e., the unit is checked only by random inspection, the total expected cost is (Problem 1 in Sect. 4.5)

$$C(\infty) \equiv \lim_{T \to \infty} C(T)$$

$$= c_R \int_0^\infty [1 + M(t)] \, dF(t) + c_D \left(\int_0^\infty F(t)\overline{G}(t) \, dt \right.$$

$$+ \int_0^\infty \left[\int_0^\infty [F(t+x) - F(x)]\overline{G}(t) \, dt \right\} dM(x) \right)$$

$$= \left(c_R + \frac{c_D}{\theta} \right) \left[1 + \int_0^\infty \overline{F}(t) \, dM(t) \right] - c_D \mu. \tag{4.4}$$

Next, when $G(t) = 1 - e^{-\theta t}$ $(0 < \theta < \infty)$, i.e., $M(t) = \theta t$, the total expected cost in (4.3) is (Problem 2 in Sect. 4.5)

$$C(T) = c_T \sum_{k=0}^\infty \overline{F}(kT) + c_R \theta \mu + \left(c_R - c_T + \frac{c_D}{\theta} \right)$$

$$\times \sum_{k=0}^\infty \int_{kT}^{(k+1)T} \left\{ 1 - e^{-\theta[(k+1)T - t]} \right\} dF(t). \tag{4.5}$$

We find an optimum checking time T^* which minimizes $C(T)$. Differentiating $C(T)$ with respect to T and setting it equal to zero,

$$\frac{\sum_{k=0}^\infty (k+1) \int_{kT}^{(k+1)T} \theta e^{-\theta[(k+1)T - t]} \, dF(t)}{\sum_{k=1}^\infty k f(kT)} - (1 - e^{-\theta T}) = \frac{c_T}{c_R - c_T + c_D/\theta} \tag{4.6}$$

for $c_R + c_D/\theta > c_T$. This is a necessary condition that an optimum T^* minimizes $C(T)$.

In addition, when $F(t) = 1 - e^{-\lambda t}$ for $\lambda < \theta$, the total expected cost $C(T)$ in (4.5) is

$$C(T) = \frac{c_T}{1 - e^{-\lambda T}} + \frac{c_R \theta}{\lambda} + \left(c_R - c_T + \frac{c_D}{\theta} \right) \left(1 - \frac{\lambda}{\theta - \lambda} \frac{e^{-\lambda T} - e^{-\theta T}}{1 - e^{-\lambda T}} \right). \tag{4.7}$$

Clearly,

$$C(0) \equiv \lim_{T \to 0} C(T) = \infty,$$

$$C(\infty) \equiv \lim_{T \to \infty} C(T) = c_R \left(\frac{\theta}{\lambda} + 1 \right) + \frac{c_D}{\theta}. \tag{4.8}$$

Equation (4.6) is simplified as

$$\frac{\theta}{\theta - \lambda}[1 - e^{-(\theta-\lambda)T}] - (1 - e^{-\theta T}) = \frac{c_T}{c_R - c_T + c_D/\theta}, \qquad (4.9)$$

whose left-hand side increases strictly from 0 to $\lambda/(\theta-\lambda)$. In particular, when $\theta \to 0$, i.e., $1/\theta \to \infty$, (4.9) becomes

$$\frac{1}{\lambda}(e^{\lambda T} - 1) - T = \frac{c_T}{c_D},$$

which agrees with (4.19) in Sect. 4.1.1 and (8.5) of [2, p. 204] for periodic inspection with only checking time T.

Therefore, if $c_R + c_D/\theta > (\theta/\lambda)c_T$, then there exists a finite and unique T^* ($0 < T^* < \infty$) which satisfies (4.9). The physical meaning of the condition $c_R + c_D/\theta > (1/\lambda)/(c_T/\theta)$ is that the total of checking cost and downtime cost of the mean interval between random checks is higher than the periodic cost for the expected number of random checks until replacement. Conversely, if $c_R + c_D/\theta \le (\theta/\lambda)c_T$, then $T^* = \infty$, i.e., periodic inspection is not needed and the expected cost is given in (4.8). If $\lambda > \theta$ and $c_R + c_D/\theta > c_T$, then the left-hand side of (4.9) also increases strictly with T from 0 to ∞, and hence, there exists a finite and unique T^* which satisfies (4.9). Note that if $\lambda = \theta$, then (4.9) becomes

$$\theta T - (1 - e^{-\theta T}) = \frac{c_T}{c_R - c_T + c_D/\theta},$$

whose left-hand side also increases from 0 to ∞. If $c_R + c_D/\theta \le c_T$, then a finite T^* does not exist in any cases, i.e., $T^* = \infty$.

Example 4.1 (*Periodic inspection for Weibull failure time*) Suppose that the failure time has a Weibull distribution and the working time is exponential, i.e., $F(t) = 1 - \exp(-\lambda t^\alpha)$ ($\alpha \ge 1$) and $G(t) = 1 - e^{-\theta t}$. Then, from (4.6), an optimum checking time T^* satisfies

$$\frac{\sum_{k=0}^\infty (k+1) \int_{kT}^{(k+1)T} \theta e^{-\theta[(k+1)T-t]}\lambda \alpha t^{\alpha-1} e^{-\lambda t^\alpha}\, dt}{\sum_{k=0}^\infty k\lambda\alpha(kT)^{\alpha-1}e^{-\lambda(kT)^\alpha}} - (1 - e^{-\theta T})$$

$$= \frac{c_T}{c_R - c_T + c_D/\theta}. \qquad (4.10)$$

In case of $\alpha = 1$, (4.10) agrees with (4.9). When $1/\theta = \infty$, (4.10) becomes

$$\frac{\sum_{k=0}^\infty e^{-\lambda(kT)^\alpha}}{\sum_{k=1}^\infty k\lambda\alpha(kT)^{\alpha-1}e^{-\lambda(kT)^\alpha}} - T = \frac{c_T}{c_D}, \qquad (4.11)$$

which corresponds to periodic inspection with Weibull failure time [2, p. 204].

Table 4.1 Optimum T^* and its cost rate when $1/\lambda = 100$, $c_T/c_D = 2$ and $c_R/c_D = 1$

$1/\theta$	$\alpha = 1$		$\alpha = 2$		$\alpha = 3$	
	T^*	$C(T^*)/c_D$	T^*	$C(T^*)/c_D$	T^*	$C(T^*)/c_D$
1	∞	102.000	∞	10.862	∞	6.144
2	∞	53.000	23.680	7.432	7.017	4.854
3	∞	37.333	16.699	6.803	6.512	4.613
4	∞	30.000	14.011	6.748	6.303	4.551
5	∞	26.000	12.264	6.783	6.187	4.541
10	∞	21.000	8.081	6.914	5.969	4.589
15	49.941	22.165	7.183	6.937	5.898	4.630
20	32.240	22.210	6.819	6.945	5.861	4.757
50	22.568	21.799	6.266	6.953	5.794	4.716
∞	19.355	21.487	5.954	5.966	5.748	4.771

Table 4.1 presents optimum T^* and its expected cost $C(T^*)/c_D$ for α and $1/\theta$ when $1/\lambda = 100$, $c_T/c_D = 2$ and $c_R/c_D = 1$. When $\alpha = 1$, if $1/\theta \leq 10$ then $T^* = \infty$. This indicates that T^* decreases with $1/\theta$ and α. However, if the mean working time $1/\theta$ exceeds a threshold level, then optimum values vary little for given α. Thus, it would be sufficient to check the unit at the smallest T^* for large $1/\theta$ which satisfies (4.11). It is of great interest that $C(T^*)$ has no monotonous property for $1/\theta$. This suggests that there might exist a combined inspection policy in which the unit is checked at periodic times kT and at Nth working time, which will be discussed in Sect. 4.2. □

Furthermore, we consider the inspection policy in which failures are detected only at periodic times kT. Then, because the inspection has no relation to any working time, this corresponds to the periodic inspection policy [2, p. 202]. However, it would be appropriate to assume in this model that c_D is the cost of the number of jobs for which a failed unit has worked before its detection, being no downtime cost per unit of time. Then, the total expected cost until replacement is

$$\tilde{C}(T) = \sum_{k=0}^{\infty} \int_{kT}^{(k+1)T} \{(k+1)c_T + c_D[M((k+1)T) - M(t)]\} \, dF(t)$$

$$= \sum_{k=0}^{\infty} \{c_T + c_D[M((k+1)T) - M(kT)]\}\overline{F}(kT) - c_D \int_0^{\infty} M(t) \, dF(t).$$

$$(4.12)$$

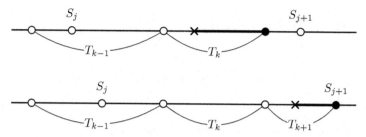

○ Check at successive and random times ✕ Failure time ● Detection of failure

Fig. 4.2 Process of random and sequential inspections

In particular, when $G(t) = 1 - e^{-\theta t}$, i.e., $M(t) = \theta t$,

$$\tilde{C}(T) = (c_T + c_D\theta T) \sum_{k=0}^{\infty} \overline{F}(kT) - c_D\theta\mu, \tag{4.13}$$

which agrees with the expected cost for standard inspection [2, p. 203] when $\theta = 1$.

4.1.1 Sequential Inspection

Suppose that the unit is checked at working times S_j $(j = 1, 2, \ldots)$ and also at successive times T_k $(k = 1, 2, \ldots)$, where $S_0 = T_0 = 0$ in Fig. 4.2. Then, by the similar method of obtaining (4.3), the total expected cost is

$$C(\mathbf{T}) = c_T \sum_{k=0}^{\infty} \overline{F}(T_k) + c_R \int_0^{\infty} M(t)\,\mathrm{d}F(t)$$

$$- (c_T - c_R)\sum_{k=0}^{\infty} \int_{T_k}^{T_{k+1}} \left\{ G(T_{k+1}) - G(t) + \int_0^t [G(T_{k+1}-x) - G(t-x)]\,\mathrm{d}M(x) \right\}\mathrm{d}F(t)$$

$$+ c_D \sum_{k=0}^{\infty} \int_{T_k}^{T_{k+1}} \left(\int_t^{T_{k+1}} \overline{G}(y)\,\mathrm{d}y + \left\{ \int_0^t \left[\int_{t-x}^{T_{k+1}-x} \overline{G}(y)\,\mathrm{d}y \right]\mathrm{d}M(x) \right\} \right)\mathrm{d}F(t), \tag{4.14}$$

where $\mathbf{T} \equiv (T_1, T_2, \ldots)$. In particular, when $G(t) = 1 - e^{-\theta t}$,

$$C(\mathbf{T}) = c_T \sum_{k=0}^{\infty} \overline{F}(T_k) + c_R \theta \mu$$

$$+ \left(c_R - c_T + \frac{c_D}{\theta} \right) \sum_{k=0}^{\infty} \int_{T_k}^{T_{k+1}} [1 - e^{-\theta(T_{k+1}-t)}] \, dF(t). \qquad (4.15)$$

Differentiating $C(\mathbf{T})$ with T_k and setting it equal to zero,

$$1 - e^{-\theta(T_{k+1}-T_k)} = \frac{\int_{T_{k-1}}^{T_k} \theta e^{-\theta(T_k-t)} \, dF(t)}{f(T_k)} - \frac{c_T}{c_R - c_T + c_D/\theta} \quad (k = 1, 2, \ldots). \tag{4.16}$$

When $1/\theta = \infty$, i.e., $\theta \to 0$, (4.16) becomes

$$T_{k+1} - T_k = \frac{F(T_k) - F(T_{k-1})}{f(T_k)} - \frac{c_T}{c_D}, \tag{4.17}$$

which corresponds to the sequential inspection [1, p. 110], [2, p. 203]. Therefore, by using Algorithm [1, p. 112], [2, p. 203], we can compute an optimum inspection schedule which satisfies (4.16).

Example 4.2 (*Inspection for Weibull failure time*) Suppose that the failure time has a Weibull distribution $[1 - \exp(-\lambda t^2)]$. Then, (4.16) is

$$1 - e^{-\theta(T_{k+1}-T_k)} = \frac{\int_{T_{k-1}}^{T_k} \theta e^{-\theta(T_k-t)} t e^{-\lambda t^2} \, dt}{T_k e^{-\lambda(T_k)^2}} - \frac{c_T}{c_R - c_T + c_D/\theta}.$$

When $1/\theta \to \infty$, from (4.17),

$$T_{k+1} - T_k = \frac{e^{-\lambda(T_{k-1})^2} - e^{-\lambda(T_k)^2}}{2\lambda T_k e^{-\lambda(T_k)^2}} - \frac{c_T}{c_D}.$$

Table 4.2 presents optimum T_k^* ($k = 1, 2, \ldots, 10$) for $1/\theta$ when $1/\lambda = 100$, $c_T/c_D = 2$ and $c_R/c_D = 1$. Note that the mean failure time is $50\sqrt{\pi} = 88.6$. This indicates that T_k^* decreases slowly with $1/\theta$, however, varies a little for $1/\theta$ and increases gradually with k. Compared to Table 4.1 when $\alpha = 2$, it is of interest that $T_1^* > T^* > T_2^* - T_1^*$ for the same $1/\theta$. $\qquad \Box$

Table 4.2 Optimum T_k^* when $F(t) = 1 - e^{-(t/10)^2}$, $c_T/c_D = 2$ and $c_R/c_D = 1$

k	$1/\theta$		
	10	50	∞
1	9.85	8.63	8.36
2	14.25	12.75	12.42
3	17.81	16.12	15.75
4	20.92	19.09	18.68
5	23.75	21.79	21.35
6	26.36	24.29	23.83
7	28.80	26.63	26.16
8	31.11	28.84	28.36
9	33.30	30.90	30.46
10	35.34	32.70	32.48

4.1.2 Comparison of Periodic and Random Inspections

Suppose that the failure time has an exponential distribution $(1 - e^{-\lambda t})$ $(0 < \lambda < \infty)$. The unit is checked at periodic times kT $(k = 1, 2, \ldots)$, and its failure is detected at the next check. Then, the total expected cost until failure detection is, from (4.7) as $\theta \to 0$ [2, p. 204],

$$C_P(T) = \frac{c_T + c_D T}{1 - e^{-\lambda T}} - \frac{c_D}{\lambda}. \tag{4.18}$$

An optimum T_S^* that minimizes $C_P(T)$ is given by a finite and unique solution of the equation

$$e^{\lambda T} - (1 + \lambda T) = \frac{c_T}{c_D/\lambda}, \tag{4.19}$$

and the resulting cost is

$$\frac{C_P(T_S^*)}{c_D/\lambda} = e^{\lambda T_S^*} - 1. \tag{4.20}$$

Next, the unit is checked at random working times S_j $(j = 1, 2, \ldots)$, where $S_0 \equiv 0$ and $Y_j = S_j - S_{j-1}$ $(j = 1, 2, \ldots)$ have an independent and exponential distribution $\Pr\{Y_j \leq t\} = 1 - e^{-\theta t}$. Then, the total expected cost until failure detection is, from (4.8),

$$C_R(\theta) = c_R \left(\frac{\theta}{\lambda} + 1 \right) + \frac{c_D}{\theta}. \tag{4.21}$$

An optimum θ^* that minimizes $C_R(\theta)$ is easily given by

Table 4.3 Optimum T_S^*, $1/\theta^*$ and their cost rates when $c_T = c_R$ and $\lambda = 1$

c_T/c_D	T_S^*	$C_P(T_S^*)/c_D$	$1/\theta^*$	$C_R(\theta^*)/c_D$
0.001	0.0444	0.0454	0.0316	0.0642
0.002	0.0626	0.0646	0.0447	0.0914
0.005	0.0984	0.1034	0.0707	0.1464
0.010	0.1382	0.1482	0.1000	0.2100
0.020	0.1935	0.2135	0.1414	0.3028
0.050	0.3004	0.3504	0.2236	0.4972
0.100	0.4162	0.5162	0.3162	0.7324
0.200	0.5722	0.7722	0.4472	1.0944
0.500	0.8577	1.3577	0.7071	1.9142
1.000	1.1462	2.1462	1.0000	3.0000

$$\frac{1}{\theta^*} = \sqrt{\frac{c_R}{c_D \lambda}}, \tag{4.22}$$

and the resulting cost is

$$\frac{C_R(\theta^*)}{c_D/\lambda} = \left(\frac{\lambda}{\theta^*}\right)^2 + 2\left(\frac{\lambda}{\theta^*}\right). \tag{4.23}$$

Example 4.3 (*Comparison of periodic and random inspections*) Table 4.3 presents optimum T_S^*, $1/\theta^*$, and their resulting costs $C_P(T_S^*)/c_D$ and $C_R(\theta^*)/c_D$ for c_T/c_D when $\lambda = 1$ and $c_T = c_R$. Both T_S^* and $1/\theta^*$ increase with c_T. This indicates as estimated previously that $T_S^* > 1/\theta^*$ and $C_P(T_S^*) < C_R(\theta^*)$, i.e., the periodic checking time is greater than the random one, and periodic inspection is better than random one numerically. This shows that if a random inspection cost c_R is the half of c_T, both expected costs of periodic and random inspections are almost the same. For example, $C_P(T_S^*)/c_D = 0.0646$ for $c_T/c_D = 0.002$ and $C_R(\theta^*)/c_D = 0.0642$ when $c_R/c_D = 0.001$. □

We compare periodic and random inspections theoretically when $c_T = c_R$. It is assumed for the simplicity of notations that $\lambda = 1$ and $c \equiv \lambda c_T/c_D \leq 1$ because the downtime cost for the mean failure time $1/\lambda$ would be much higher than that one checking cost for most inspection models. When $c = 1$, $T_S^* = 1.1462$ and $1/\theta^* = 1.0$. Thus, it is easily noted that $0 < T_S^* \leq 1.1462$ and $0 < 1/\theta^* \leq 1.0$. From (4.19) and (4.22), a solution of the equation

$$Q(T) = e^T - (1 + T + T^2) = 0$$

is $T = 1.79 > 1.1462$, which follows that $Q(T) < 0$ for $0 < T < 1.79$. Thus, $0 < 1/\theta^* < T_S^* \leq 1.1462$.

Next, prove that $2/\theta^* > T_S^*$. From (4.19),

$$c = e^T - (1+T) > \frac{T^2}{2}, \quad \text{i.e.,} \quad T_S^* < \sqrt{2c},$$

and furthermore,

$$\frac{2}{\theta^*} = 2\sqrt{c} > \sqrt{2c} > T_S^*.$$

So that,

$$\frac{1}{\theta^*} < T_S^* < \frac{2}{\theta^*}.$$

In addition, from (4.20) and (4.23),

$$\frac{C_R(\theta^*) - C_P(T_S^*)}{c_D} = \left(\frac{1}{\theta^*}\right)^2 + \frac{2}{\theta^*} - e^{T_S^*} + 1 > \frac{2}{\theta^*} - T_S^* > 0.$$

From the above results, $T_S^* > 1/\theta^*$ and $C_P(T_S^*) < C_R(\theta^*)$, i.e., periodic inspection is better than random one and the optimum interval T_S^* is greater than $1/\theta^*$.

It has been assumed until now that both checking costs for periodic and random inspections are the same. Usually, the cost for random check would be lower than that for periodic one because the unit is checked at random times. We compute a random checking cost $\widehat{c}_R$ when the expected costs of two inspections are the same one. We compute $1/\widehat{\theta}$ for c from Table 4.1 when

$$C_P(T_S^*) = e^{T_S^*} - 1 = \left(\frac{1}{\widehat{\theta}}\right)^2 + \frac{2}{\widehat{\theta}},$$

and using $\widehat{\theta}$, we obtain

$$\frac{\widehat{c}_R}{c_D} = \left(\frac{1}{\widehat{\theta}}\right)^2.$$

Example 4.4 (Random checking cost) Table 4.4 presents $1/\widehat{\theta}, \widehat{c}_R/c_D$ and $\widehat{c}_R/c_T$ for c_T/c_D, and indicates that the checking cost $\widehat{c}_R$ for random inspection is a little higher than the half of c_T. It is noted from (4.19) and (4.22) that $T_S^* \to 0$ and $1/\theta^* \to 0$ as $c_T \to 0$. Thus, from (4.20) and (4.23),

$$\lim_{T \to 0} \frac{e^{\lambda T} - 1}{(\lambda T)^2 + 2\lambda T} = \frac{1}{2}.$$

This shows that if $c_T \to 0$, then $C_P(T_S^*) \to C_R(\theta)/2$, i.e., as $c_T \to 0$, the expected cost of periodic inspection is the half of that of random one. Therefore, it would be estimated that if $c_T \to 0$ and $c_R/c_T = 0.5$, then both expected costs of periodic

Table 4.4 Values of $1/\widehat{\theta}$, $\widehat{c}_R/c_D$ and $\widehat{c}_R/c_T$

c_T/c_D	$1/\widehat{\theta}$	$\widehat{c}_R/c_D$	$\widehat{c}_R/c_T$
0.001	0.0224	0.0005	0.5039
0.002	0.0318	0.0010	0.5054
0.005	0.0504	0.0025	0.5086
0.010	0.0715	0.0051	0.5118
0.020	0.1016	0.0103	0.5160
0.050	0.1621	0.0263	0.5253
0.100	0.2313	0.0535	0.5352
0.200	0.3312	0.1097	0.5485
0.500	0.5355	0.2868	0.5735
1.000	0.7738	0.5987	0.5987

and random inspections would be the same, as shown in Table 4.3 (Problem 3 in Sect. 4.5).   □

4.2 Random Inspection

Suppose that the unit is checked at every Nth ($N = 1, 2, \ldots$) working times S_{jN} ($j = 1, 2, \ldots$), i.e., at the jNth number of works, and also at periodic times kT ($k = 1, 2, \ldots$), whichever occurs first. Then, the total expected cost until failure detection is, by replacing formally $G(t)$ and $M(t)$ with $G^{(N)}(t)$ and $M^{(N)}(t) \equiv \sum_{j=1}^{\infty} G^{(jN)}(t)$ ($N = 1, 2, \ldots$) in (4.3), respectively,

$$
C(T, N) = c_T \sum_{k=0}^{\infty} \overline{F}(kT) + c_R \int_0^{\infty} M^{(N)}(t)\, dF(t)
$$

$$
- (c_T - c_R) \sum_{k=0}^{\infty} \int_{kT}^{(k+1)T} \left\{ G^{(N)}((k+1)T) - G^{(N)}(t) \right.
$$

$$
+ \int_0^t [G^{(N)}((k+1)T - x) - G^{(N)}(t - x)]\, dM^{(N)}(x) \left.\right\}\, dF(t)
$$

$$
+ c_D \sum_{k=0}^{\infty} \int_{kT}^{(k+1)T} \left(\int_t^{(k+1)T} [1 - G^{(N)}(y)]\, dy \right.
$$

$$
+ \int_0^t \left\{ \int_{t-x}^{(k+1)T-x} [1 - G^{(N)}(y)]\, dy \right\}\, dM^{(N)}(x) \left.\right)\, dF(t), \qquad (4.24)
$$

where $M^{(1)}(t) \equiv M(t)$. In general, it is very difficult to derive analytically both optimum T^* and N^* which minimize $C(T, N)$.

In particular, when $T = \infty$, i.e., the unit is checked only at every Nth working times, the total expected cost is

$$
\begin{aligned}
C_R(N) &\equiv \lim_{T \to \infty} C(T, N) \\
&= c_T + c_R \int_0^\infty M^{(N)}(t) \, dF(t) - (c_T - c_R) \\
&\quad + \frac{N c_D}{\theta} \left[\int_0^\infty \overline{F}(t) \, dM^{(N)}(t) + 1 \right] - c_D \mu \\
&= \left(c_R + \frac{N c_D}{\theta} \right) \left[1 + \int_0^\infty \overline{F}(t) \, dM^{(N)}(t) \right] - c_D \mu \quad (N = 1, 2, \ldots).
\end{aligned}
$$

$$(4.25)$$

This is also obtained easily from (4.4), by replacing $M(t)$ and $1/\theta$ with $M^{(N)}(t)$ and N/θ formally. In addition, when $F(t) = 1 - e^{-\lambda t}$ $(0 < \lambda < \infty)$,

$$
\int_0^\infty e^{-\lambda t} \, dM^{(N)}(t) = \sum_{j=1}^\infty \int_0^\infty e^{-\lambda t} \, dG^{(jN)}(t) = \frac{[G^*(\lambda)]^N}{1 - [G^*(\lambda)]^N},
$$

where $G^*(s)$ is the LS transform of $G(t)$, i.e., $G^*(s) \equiv \int_0^\infty e^{-st} \, dG(t)$ for $Re(s) > 0$. In this case, the expected cost in (4.25) is

$$
C_R(N) = \frac{c_R + N c_D/\theta}{1 - [G^*(\lambda)]^N} - \frac{c_D}{\lambda}. \tag{4.26}
$$

We find an optimum number N^* which minimizes $C_R(N)$. From the inequality $C_R(N+1) - C_R(N) \geq 0$,

$$
\sum_{j=1}^N \left[\frac{1}{G^*(\lambda)} \right]^j - N \geq \frac{c_R}{c_D/\theta}, \tag{4.27}
$$

whose left-hand side increases strictly from $1/G^*(\lambda) - 1$ to ∞. Therefore, there exists a finite and unique minimum N^* $(1 \leq N^* < \infty)$ which satisfies (4.27).

Example 4.5 (Random inspection for exponential failure and working times) When $G(t) = 1 - e^{-\theta t}$, i.e., $G^*(\lambda) = \theta/(\lambda + \theta)$, the total expected cost is, from (4.26),

Table 4.5 Optimum N^* and its cost rate when $1/\lambda = 100$

$\dfrac{1}{\theta}$	$c_R/c_D = 1$		$c_R/c_D = 5$	
	N^*	$C_R(N^*)/c_D$	N^*	$C_R(N^*)/c_D$
1	14	15.352	30	35.618
2	7	15.884	15	36.195
3	5	16.456	10	36.769
4	4	17.083	8	37.388
5	3	17.507	6	37.912
10	1	21.000	3	40.740
15	1	22.667	2	43.527
20	1	26.000	2	47.273
25	1	30.000	1	50.000

$$C_R(N) = \frac{c_R + Nc_D/\theta}{1 - [\theta/(\lambda+\theta)]^N} - \frac{c_D}{\lambda}, \tag{4.28}$$

and from (4.27), an optimum N^* satisfies

$$\sum_{j=1}^{N}\left(1+\frac{\lambda}{\theta}\right)^j - N \geq \frac{c_R}{c_D/\theta},$$

i.e.,

$$\left(1+\frac{\lambda}{\theta}\right)^{N+1} - (N+1)\frac{\lambda}{\theta} - 1 \geq \frac{c_R}{c_D/\lambda}, \tag{4.29}$$

whose left-hand increases strictly with N from $(\lambda/\theta)^2$ to ∞. Thus, there exists a finite and unique minimum N^* ($1 \leq N^* < \infty$) which satisfies (4.29). If $1/\theta \geq \sqrt{c_R/(\lambda c_D)}$, then $N^* = 1$. It can be shown that because the left-hand of (4.29) increases strictly with $1/\theta$ from 0 to ∞, N^* decreases with $1/\theta$ from ∞ to 1.

Table 4.5 presents optimum N^* and its expected cost $C_R(N^*)/c_D$ for $1/\theta$ and c_R/c_D when $1/\lambda = 100$. This indicates that optimum N^* decreases with $1/\theta$ and increases with c_R/c_D, however, N^*/θ is almost the same for small $1/\theta$. Compared to Table 4.1 when $\alpha = 1$, if $1/\theta$ is small, then $C_R(N^*)$ is less than $C(T^*)$. Note that when $1/\theta = 10$, $C(T^*) = C_R(N^*)$ because $N^* = 1$ and $T^* = \infty$. We should adopt periodic inspection when the mean working time $1/\theta$ is long. These tables suggest that if the working time is not so large, i.e., $1/\theta < 10$, random inspection is better than periodic one. □

Example 4.6 (Inspection number for Weibull failure and exponential working times)
Suppose that the failure time has a Weibull distribution $F(t) = 1 - \exp(-\lambda t^\alpha)$ ($\alpha \geq 1$), $\mu = \Gamma(1+1/\alpha)/\lambda^{1/\alpha}$, and $G(t) = 1 - e^{-\theta t}$. In this case, because the renewal density is $m^{(N)}(t) \equiv dM^{(N)}(t)/dt$, from [1, p. 57], [10, p. 52],

Table 4.6 Optimum N^* when $1/\lambda = 100$ and $c_R/c_D = 5$

$1/\theta$	$\alpha = 1$	$\alpha = 2$	$\alpha = 3$
1	30	9	6
2	15	5	3
3	10	3	2
4	8	2	2
5	6	2	2
10	3	1	1
15	2	1	1
20	2	1	1
25	1	1	1

$$m^{(N)}(t) = \sum_{j=1}^{\infty} \frac{\theta(\theta t)^{Nj-1}}{(Nj-1)!} e^{-\theta t},$$

the expected cost in (4.25) is

$$C_R(N) = \left(c_R + \frac{Nc_D}{\theta}\right)\left[1 + \sum_{j=1}^{\infty}\int_0^{\infty} e^{-\lambda t^\alpha} \frac{\theta(\theta t)^{Nj-1}}{(Nj-1)!} e^{-\theta t}\, dt\right] - c_D\mu$$

$$(N = 1, 2, \ldots). \tag{4.30}$$

Table 4.6 presents optimum N^* for α and $1/\theta$ when $1/\lambda = 100$ and $c_R/c_D = 5$. When $\alpha = 1$, N^* is equal to Table 4.5 when $c_R/c_D = 5$. Because the failure rate $h(t) = \lambda\alpha t^{\alpha-1}$ ($\alpha > 1$) increases rapidly, N^* becomes much smaller than that for $\alpha = 1$ when the working time is small, and N^*/θ is almost constant, i.e., $N^*/\theta \approx 30, 9, 6$ for $\alpha = 1, 2, 3$, respectively. On the other hand, when the working time is very large, the unit should be checked at every working times, i.e., $N^* = 1$. ☐

4.3 Modified Random Inspection Policies

As modified random inspection policies, we propose the following three random policies of inspection first, last, and overtime, and derive their optimum policies which minimize the expected costs, as shown in Chaps. 2 and 3.

4.3.1 Inspection First

Suppose that the unit is checked at a planned time T ($0 < T \leq \infty$) or at a random working time Y_j ($j = 1, 2, \ldots$), whichever occurs first. That is, the unit is checked

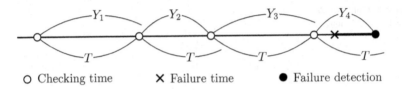

Fig. 4.3 Process of inspection first

at interval times $Z_j \equiv \min\{T, Y_j\}$ $(j = 1, 2, \ldots)$ in Fig. 4.3, and Y_j has an identical distribution $G(t) \equiv \Pr\{Y_j \leq t\}$. In this case, Z_j forms a renewal process with an interarrival distribution $\Pr\{Z_j \leq t\} = G(t)$ for $t < T$, 1 for $t \geq T$.

It is assumed that the failure time has an exponential distribution $F(t) = 1 - e^{-\lambda t}$ $(0 < \lambda < \infty)$. Then, the probability that the unit does not fail and is checked at time T is

$$\overline{G}(T)\overline{F}(T), \tag{4.31}$$

the probability that it does not fail and is checked at time Y_j is

$$\int_0^T \overline{F}(t) \, dG(t), \tag{4.32}$$

the probability that it fails and its failure is detected at time T is

$$\overline{G}(T)F(T), \tag{4.33}$$

and the probability that it fails and its failure is detected at time Y_j is

$$\int_0^T F(t) \, dG(t), \tag{4.34}$$

where $(4.31)+(4.32)+(4.33)+(4.34)=1$.

From (4.31) to (4.34), the mean downtime l_D from a failure to its detection is given by a renewal equation

$$l_D \equiv \left[\overline{G}(T)\overline{F}(T) + \int_0^T \overline{F}(t) \, dG(t) \right] l_D$$

$$+ \int_0^T (T - t)\overline{G}(T) \, dF(t) + \int_0^T \left[\int_0^t (t - u) \, dF(u) \right] dG(t).$$

By solving the above renewal equation and arranging it,

$$l_D = \frac{\int_0^T \overline{G}(t) F(t) \, dt}{\int_0^T \overline{G}(t) \, dF(t)}. \tag{4.35}$$

In a similar way, the expected number M_T of checks at time T until failure detection is given by a renewal equation

$$M_T = (1 + M_T)\overline{G}(T)\overline{F}(T) + M_T \int_0^T \overline{F}(t) \, dG(t) + \overline{G}(T)F(T),$$

i.e.,

$$M_T = \frac{\overline{G}(T)}{\int_0^T \overline{G}(t) \, dF(t)}. \tag{4.36}$$

The expected number M_R of checks at time Y_j until failure detection is given by a renewal equation

$$M_R = (1 + M_R) \int_0^T \overline{F}(t) \, dG(t) + M_R \overline{G}(T)\overline{F}(T) + \int_0^T F(t) \, dG(t),$$

i.e.,

$$M_R = \frac{G(T)}{\int_0^T \overline{G}(t) \, dF(t)}. \tag{4.37}$$

Therefore, the total expected cost until failure detection is

$$\begin{aligned} C_F(T) &= c_T M_T + c_R M_R + c_D l_D \\ &= \frac{c_T \overline{G}(T) + c_R G(T) + c_D \int_0^T \overline{G}(t) F(t) \, dt}{\int_0^T \overline{G}(t) \, dF(t)}, \end{aligned} \tag{4.38}$$

where c_T, c_R and c_D are given in (4.3).
When $G(t) = 1 - e^{-\theta t} \ (0 < \theta < \infty)$,

$$C_F(T) = \frac{c_T + (c_R - c_T + c_D/\theta)(1 - e^{-\theta T})}{[\lambda/(\theta + \lambda)][1 - e^{-(\theta + \lambda)T}]} - \frac{c_D}{\lambda}. \tag{4.39}$$

In particular,

$$\lim_{\theta \to 0} C_F(T) = \frac{c_T + c_D T}{1 - e^{-\lambda T}} - \frac{c_D}{\lambda},$$

which agrees with (4.18), and

$$\lim_{T \to \infty} C_F(T) = c_R \left(\frac{\theta}{\lambda} + 1 \right) + \frac{c_D}{\theta},$$

which agrees with (4.8). This policy includes periodic and random inspections discussed in Sect. 4.1. We find an optimum T_F^* which minimizes $C_F(T)$ in (4.39) for $c_R + c_D/\theta > c_T$. Differentiating $C_F(T)$ with respect to T and setting it equal to zero,

$$\frac{\theta}{\theta + \lambda}(e^{\lambda T} - 1) - \frac{\lambda}{\theta + \lambda}(1 - e^{-\theta T}) = \frac{c_T}{c_R - c_T + c_D/\theta}, \qquad (4.40)$$

whose left-hand side increases strictly from 0 to ∞. Thus, there exists a finite and unique T_F^* $(0 < T_F^* < \infty)$ which satisfies (4.40), and the resulting cost rate is

$$\frac{\lambda C_F(T_F^*)}{\theta(c_R - c_T) + c_D} = e^{\lambda T_F^*} - \frac{c_D}{\theta(c_R - c_T) + c_D}. \qquad (4.41)$$

When $c_R = c_T$, (4.40) is

$$\frac{1}{\theta + \lambda}(e^{\lambda T} - 1) - \frac{\lambda}{\theta(\theta + \lambda)}(1 - e^{-\theta T}) = \frac{c_T}{c_D},$$

whose left-hand side decreases with θ, and T_F^* increases with θ from T_S^* given in (4.19) to ∞ (Problem 4 in Sect. 4.5).

4.3.2 Inspection Last

Suppose that the unit is checked at a planned time T $(0 \le T < \infty)$ or at a random working time Y_j $(j = 1, 2, \ldots)$, whichever occurs last. That is, the unit is checked at interval times $\tilde{Z}_j \equiv \max\{T, Y_j\}$ $(j = 1, 2, \ldots)$ with $G(t) \equiv \Pr\{Y_j \le t\}$ in Fig. 4.4. In this case, $\tilde{Z}_j$ forms a renewal process with an interarrival distribution $\Pr\{\tilde{Z}_j \le t\} = 0$ for $t < T$, and $G(t)$ for $t \ge T$.

It is assumed that the failure time has an exponential distribution $F(t) = 1 - e^{-\lambda t}$. Then, the probability that the unit does not fail and is checked at time T is

$$G(T)\overline{F}(T), \qquad (4.42)$$

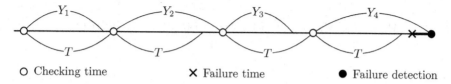

Fig. 4.4 Process of inspection last

the probability that it does not fail and is checked at time Y_j is

$$\int_T^\infty \overline{F}(t)\,dG(t), \tag{4.43}$$

the probability that it fails and its failure is detected at time T is

$$G(T)F(T), \tag{4.44}$$

and the probability that it fails and its failure is detected at time Y_j is

$$\int_T^\infty F(t)\,dG(t), \tag{4.45}$$

where $(4.42)+(4.43)+(4.44)+(4.45)=1$. From (4.42) to (4.45), the mean down-time l_D from a failure to its detection is given by a renewal equation

$$l_D = \left[G(T)\overline{F}(T) + \int_T^\infty \overline{F}(t)\,dG(t) \right] l_D$$

$$+ \int_0^T (T-t)G(T)\,dF(t) + \int_T^\infty \left[\int_0^t (t-u)\,dF(u) \right] dG(t).$$

By solving the above renewal equation,

$$l_D = \frac{\int_0^T F(t)\,dt + \int_T^\infty \overline{G}(t)F(t)\,dt}{1 - \int_T^\infty G(t)\,dF(t)}. \tag{4.46}$$

In a similar way, the expected number M_T of checks at time T until failure detection is given by a renewal equation

$$M_T = (1 + M_T)G(T)\overline{F}(T) + M_T \int_T^\infty \overline{F}(t)\,dG(t) + G(T)F(T),$$

i.e.,

$$M_T = \frac{G(T)}{1 - \int_T^\infty G(t)\,dF(t)}. \qquad (4.47)$$

The expected number M_R of checks at time Y_j until failure detection is given by a renewal equation

$$M_R = (1 + M_R)\int_T^\infty \overline{F}(t)\,dG(t) + M_R G(T)\overline{F}(T) + \int_T^\infty F(t)\,dG(t),$$

i.e.,

$$M_R = \frac{\overline{G}(T)}{1 - \int_T^\infty G(t)\,dF(t)}. \qquad (4.48)$$

Therefore, the total expected cost until failure detection is, from (4.38),

$$C_L(T) = \frac{c_T G(T) + c_R \overline{G}(T) + c_D[\int_0^T F(t)\,dt + \int_T^\infty \overline{G}(t)F(t)\,dt]}{1 - \int_T^\infty G(t)\,dF(t)}. \qquad (4.49)$$

When $G(t) = 1 - e^{-\theta t}$,

$$C_L(T) = \frac{c_T(1 - e^{-\theta T}) + c_R e^{-\theta T} + (c_D/\theta)(\theta T + e^{-\theta T})}{1 - e^{-\lambda T} + [\lambda/(\theta + \lambda)]e^{-(\theta+\lambda)T}} - \frac{c_D}{\lambda}. \qquad (4.50)$$

In particular,

$$\lim_{\theta \to \infty} C_L(T) = \lim_{\theta \to 0} C_F(T) = C_P(T), \qquad \lim_{T \to 0} C_L(T) = \lim_{T \to \infty} C_F(T) = C_R(\theta),$$

which are given in (4.18) and (4.21), respectively. We find an optimum T_L^* which minimizes $C_L(T)$ in (4.50). Differentiating $C_L(T)$ with respect to T and setting it equal to zero,

$$\frac{c_T - c_R}{e^{\theta T} - 1}\left[\frac{\theta}{\lambda}(e^{\lambda T} - 1) + \frac{\lambda}{\theta + \lambda}(1 - e^{-\theta T}) + \frac{\theta}{\theta + \lambda}\right]$$
$$+ \frac{c_D}{\theta}\left\{\frac{\theta}{\lambda}[e^{\lambda T} - (1 + \lambda T)] - \frac{\lambda}{\theta + \lambda}e^{-\theta T}\right\} = c_T. \qquad (4.51)$$

When $c_T = c_R$, (4.51) becomes

$$\frac{\theta}{\lambda}[e^{\lambda T} - (1 + \lambda T)] - \frac{\lambda}{\theta + \lambda}e^{-\theta T} = \frac{c_T}{c_D/\theta}, \tag{4.52}$$

whose left-hand side increases strictly with T from $-\lambda/(\theta + \lambda)$ to ∞. Thus, there exists a finite and unique $\tilde{T}_L$ ($0 < \tilde{T}_L < \infty$) which satisfies (4.52). Therefore, an optimum T_L^* which minimizes $C_L(T)$ in (4.51) is $T_L^* \geq \tilde{T}_L$ for $c_R \geq c_T$ and $T_L^* < \tilde{T}_L$ for $c_R < c_T$. Clearly, $\tilde{T}_L$ decreases strictly with θ to T_S^* given in (4.19) (Problem 5 in Sect. 4.5).

4.3.3 Comparison of Inspection First and Last

We compare optimum policies for inspection first and last when $c_T = c_R$ and $0 < \theta < \infty$ and $0 < \lambda < \infty$. In this case, the expected cost of inspection first is, from (4.39),

$$C_F(T) = \frac{c_T + (c_D/\theta)(1 - e^{-\theta T})}{[\lambda/(\theta + \lambda)][1 - e^{-(\theta + \lambda)T}]} - \frac{c_D}{\lambda}, \tag{4.53}$$

an optimum T_F^* which minimizes it, is from (4.40),

$$\frac{\theta}{\theta + \lambda}(e^{\lambda T} - 1) - \frac{\lambda}{\theta + \lambda}(1 - e^{-\theta T}) = \frac{c_T}{c_D/\theta}, \tag{4.54}$$

and the resulting cost is

$$C_F(T_F^*) = \frac{c_D}{\lambda}(e^{\lambda T_F^*} - 1). \tag{4.55}$$

The expected cost of inspection last is, from (4.50),

$$C_L(T) = \frac{c_T + (c_D/\theta)(\theta T + e^{-\theta T})}{1 - e^{-\lambda T} + [\lambda/(\theta + \lambda)]e^{-(\theta + \lambda)T}} - \frac{c_D}{\lambda}, \tag{4.56}$$

an optimum T_L^* which minimizes it is given in (4.52), and the resulting cost is

$$C_L(T_L^*) = \frac{c_D}{\lambda}(e^{\lambda T_L^*} - 1). \tag{4.57}$$

By comparing (4.19) with (4.54) for $0 < T < \infty$,

$$\frac{1}{\lambda}[e^{\lambda T} - (1 + \lambda T)] > \frac{1}{\theta + \lambda}(e^{\lambda T} - 1) - \frac{\lambda}{\theta(\theta + \lambda)}(1 - e^{-\theta T}),$$

which follows that $T_F^* > T_S^*$. Similarly, by comparing (4.19) with (4.52), $T_L^* > T_S^*$. Therefore, from (4.20), (4.55) and (4.57), periodic inspection with only time T is better than both inspection first and inspection last.

Furthermore, to compare (4.52) with (4.54),

$$Q(T) \equiv \frac{\theta}{\lambda}[e^{\lambda T} - (1 + \lambda T)] - \frac{\lambda}{\theta + \lambda}e^{-\theta T}$$
$$- \frac{\theta}{\theta + \lambda}(e^{\lambda T} - 1) + \frac{\lambda}{\theta + \lambda}(1 - e^{-\theta T})$$
$$= \frac{\theta}{\lambda}[e^{\lambda T} - (1 + \lambda T)] + \frac{\lambda}{\theta + \lambda}(1 - 2e^{-\theta T}) - \frac{\theta}{\theta + \lambda}(e^{\lambda T} - 1).$$

Clearly, $Q(T)$ increases strictly with T from $-\lambda/(\theta + \lambda)$ to ∞. Thus, there exists a finite and unique T_I ($0 < T_I < \infty$) which satisfies $Q(T) = 0$. Therefore, from (4.52) and (4.54), if

$$L(T_I) \equiv \frac{\theta}{\theta + \lambda}(e^{\lambda T_I} - 1) - \frac{\lambda}{\theta + \lambda}(1 - e^{-\theta T_I}) > \frac{c_T}{c_D/\theta}, \qquad (4.58)$$

then $T_F^* < T_L^*$, and hence, inspection first is better than inspection last, and conversely, if $L(T_I) < c_T/(c_D/\theta)$, $T_L^* < T_F^*$, and hence, inspection last is better than inspection first.

Example 4.7 (Checking time for inspection first and last) Table 4.7 presents optimum T_F^* and T_L^* which satisfy (4.40) and (4.52), respectively, and T_I, $L(T_I)$ for c_T/c_D and $1/\theta$ when $1/\lambda = 1$ and $c_T = c_R$. When $1/\theta = \infty$, T_S^* agrees with that in Table 4.3. This indicates that both T_F^* and T_L^* increase with c_T/c_D. When c_T/c_D is small, i.e., $L(T_I) > c_T/c_D$, $T_F^* < T_L^*$ and inspection first is better than inspection last. Conversely, when c_T/c_D is large, i.e., $L(T_I) < c_T/c_D$, $T_L^* < T_F^*$ and inspection last is better than inspection first. Optimum T_F^* decreases with $1/\theta$ to T_S^* and T_L^* increases with $1/\theta$ from T_S^*. Furthermore, inspection first is better than inspection last as $1/\theta$ becomes larger. It is of interest that when $1/\theta = 0.5$ and $c_T/c_D = 0.100$, $T_F^* = 0.4739 < 1/\theta = 0.5 < T_L^* = 0.5161$, and both inspection times are almost the same.   □

4.3.4 Inspection Overtime

It is assumed that the unit has an exponential failure distribution $F(t) = 1 - e^{-\lambda t}$ and random working times are exponential, i.e., $G(t) = 1 - e^{-\theta t}$. Suppose that the unit is checked at the first completion of working times over time T ($0 \le T < \infty$). Such inspection procedures have continued until failure detection. This is called *inspection overtime*.

Table 4.7 Optimum T_F^*, T_L^* and T_I when $\lambda = 1$

$\frac{c_T}{c_D}$	$1/\theta = 0.1$		$1/\theta = 0.2$		$1/\theta = 0.5$		$1/\theta = \infty$
	T_F^*	T_L^*	T_F^*	T_L^*	T_F^*	T_L^*	T_S^*
0.001	0.0479	0.0939	0.0461	0.1698	0.0450	0.3746	0.0444
0.002	0.0697	0.1012	0.0660	0.1737	0.0639	0.3762	0.0626
0.005	0.1168	0.1216	0.1069	0.1850	0.1016	0.3811	0.0984
0.010	0.1764	0.1511	0.1553	0.2030	0.1446	0.3891	0.1382
0.020	0.2727	0.1993	0.2279	0.2362	0.2061	0.4048	0.1936
0.050	0.5004	0.3017	0.3859	0.3190	0.3307	0.4492	0.3004
0.100	0.7884	0.4165	0.5817	0.4239	0.4739	0.5161	0.4162
0.200	1.1939	0.5723	0.8744	0.5747	0.6787	0.6297	0.5722
0.500	1.8871	0.8577	1.4350	0.8580	1.0792	0.8785	0.8577
1.000	2.4932	1.1462	1.9742	1.1462	1.4985	1.1539	1.1462
T_I	0.1259		0.2444		0.5643		
$L(T_I)$	0.0057		0.0226		0.1400		

The probability that the unit does not fail at some checking interval is

$$\sum_{j=0}^{\infty} \int_0^T \left[\int_T^{\infty} \overline{F}(u) \, dG(u-t) \right] dG^{(j)}(t) = \frac{\theta}{\lambda + \theta} e^{-\lambda T},$$

and the probability that it fails at some interval is

$$\sum_{j=0}^{\infty} \int_0^T \left[\int_T^{\infty} F(u) \, dG(u-t) \right] dG^{(j)}(t) = 1 - \frac{\theta}{\lambda + \theta} e^{-\lambda T}.$$

Thus, the mean time from a failure to its detection is

$$\sum_{j=0}^{\infty} \int_0^T \left\{ \int_T^{\infty} \left[\int_0^u (u-x) \, dF(x) \right] dG(u-t) \right\} dG^{(j)}(t)$$

$$= T + \frac{1}{\theta} - \frac{1}{\lambda} + \frac{\theta}{\lambda(\theta + \lambda)} e^{-\lambda T}. \tag{4.59}$$

The expected number M_C of checking times until failure detection is given by a renewal equation

$$M_C = (1 + M_C) \sum_{j=0}^{\infty} \int_0^T \left[\int_T^{\infty} \overline{F}(u) \, dG(u - t) \right] dG^{(j)}(t)$$

$$+ \sum_{j=0}^{\infty} \int_0^T \left[\int_T^{\infty} F(u) \, dG(u - t) \right] dG^{(j)}(t).$$

Thus, by solving the above equation,

$$M_C = \frac{1}{\sum_{j=0}^{\infty} \int_0^T \left[\int_T^{\infty} F(u) \, dG(u - t) \right] dG^{(j)}(t)}$$

$$= \frac{1}{1 - [\theta/(\theta + \lambda)]e^{-\lambda T}}. \tag{4.60}$$

Therefore, from (4.59) and (4.60), the total expected cost until failure detection is

$$C_O(T) = \frac{c_R + c_D(T + 1/\theta)}{1 - [\theta/(\theta + \lambda)]e^{-\lambda T}} - \frac{c_D}{\lambda}, \tag{4.61}$$

where c_R = checking cost over time T and c_D is given in (4.38). Clearly,

$$C_O(\infty) \equiv \lim_{T \to \infty} C_O(T) = \infty,$$

$$C_O(0) \equiv \lim_{T \to 0} C_O(T) = c_R \left(\frac{\theta}{\lambda} + 1 \right) + \frac{c_D}{\theta}, \tag{4.62}$$

which agrees with (4.8). We find an optimum T_O^* which minimizes $C_O(T)$ in (4.61). Differentiating $C_O(T)$ with respect to T and setting it equal to zero,

$$\left(\frac{1}{\lambda} + \frac{1}{\theta} \right) (e^{\lambda T} - 1) - T = \frac{c_R}{c_D}, \tag{4.63}$$

whose left-hand side increases strictly from 0 to ∞. Thus, there exists a finite and unique T_O^* $(0 < T_O^* < \infty)$, which satisfies (4.63), and the resulting cost is

$$\frac{C_O(T_O^*)}{c_D/\lambda} = \left(1 + \frac{\lambda}{\theta} \right) e^{\lambda T_O^*} - 1. \tag{4.64}$$

Clearly, T_O^* increases with θ from 0 to T_S^*.

Compare periodic inspection in which the expected cost is given in (4.18) with inspection overtime when $c_T = c_R$. In this case, it can be easily shown that from (4.63), T_O^* decreases with $1/\theta$ from T_S^* to 0, and

$$T_O^* < T_S^* < T_O^* + \frac{1}{\theta}.$$

From (4.19) and (4.63), $T_O^* < T_S^*$. On the other hand,

$$\frac{1}{\lambda}\left[e^{\lambda(T_O^*+1/\theta)} - 1\right] - \left(T_O^* + \frac{1}{\theta}\right) > \left(\frac{1}{\lambda} + \frac{1}{\theta}\right)(e^{\lambda T_O^*} - 1) - T_O^* = \frac{c_T}{c_D},$$

which implies that $T_O^* + 1/\theta > T_S^*$. So that, comparing (4.20) with (4.64), $C_P(T_S^*) < C_O(T_O^*)$, i.e., periodic inspection is better than inspection overtime.

Furthermore, we compare T_O^* with T_F^* and T_L^* when $c_T = c_R$: From (4.54) and (4.63),

$$\left(\frac{\theta + \lambda}{\lambda}\right)(e^{\lambda T} - 1) - \theta T - \frac{\theta}{\theta + \lambda}(e^{\lambda T} - 1) + \frac{\lambda}{\theta + \lambda}(1 - e^{-\theta T})$$

$$> \frac{\lambda^2 T}{\theta + \lambda} + \frac{\lambda}{\theta + \lambda}(1 - e^{-\theta T}) > 0.$$

Thus, $T_O^* < T_F^*$. Similarly, from (4.52) and (4.63), $T_O^* < T_L^*$ (Problem 6 in Sect. 4.5). The above results are also proved easily that $T_F^* > T_S^*$, $T_L^* > T_S^*$ and $T_S^* > T_O^*$.

Next, assume that $c_R < c_T$. Then, from (4.20) and (4.64), if

$$c_T + c_D T_S^* > c_R + c_D\left(T_O^* + \frac{1}{\theta}\right),$$

then inspection overtime is better than periodic inspection.

Furthermore, we obtain $\widehat{c}_R$ in the case where $C_P(T_S^*) = C_O(T_O^*)$ for given c_T and c_D. First, we compute T_S^* from (4.19) and $C_P(T_S^*)$ from (4.20). Using T_S^* and $C_P(T_S^*)$, we obtain $\widehat{T}_O$ which satisfies

$$\left(\frac{1}{\lambda} + \frac{1}{\theta}\right)(e^{\lambda \widehat{T}_O} - 1) + \frac{1}{\theta} = T_S^* + \frac{c_T}{c_D},$$

and from (4.63),

$$\frac{\widehat{c}_R}{c_D} = T_S^* + \frac{c_T}{c_D} - \left(\widehat{T}_O + \frac{1}{\theta}\right). \tag{4.65}$$

Example 4.8 (*Inspection for exponential failure time*) Table 4.8 presents optimum T_O^* and $\widehat{c}_R/c_D$ for $1/\theta$, c_T/c_D and c_R/c_D when $F(t) = 1 - e^{-t}$. Optimum T_O^* and $\widehat{c}_R/c_D$ increase with c_T/c_D and decrease with $1/\theta$. Compared to Table 4.7, $T_O^* < T_S^* < T_O^* + 1/\theta$. This indicates that $\widehat{c}_R/c_D$ approaches to c_T/c_D as c_T/c_D becomes larger. In other words, if $c_T (= c_R)$ becomes higher, then T_O^* and T_S^* become larger, and both overtime inspection and periodic inspection are almost the same. That is, the checking cost for overtime inspection approaches to that for periodic inspection

Table 4.8 Optimum T_0^* for $c_T = c_R$ and $\widehat{c}_R$ for $c_R < c_T$ when $\lambda = 1$

$\dfrac{c_T}{c_D}$	$1/\theta = 0.01$		$1/\theta = 0.05$		$1/\theta = 0.1$	
	T_0^*	$\widehat{c}_R/c_D$	T_0^*	$\widehat{c}_R/c_D$	T_0^*	$\widehat{c}_R/c_D$
0.001	0.0355	–	0.0170	–	0.0095	–
0.002	0.0534	0.0012	0.0303	–	0.0182	–
0.005	0.0889	0.0045	0.0606	–	0.0407	–
0.010	0.1285	0.0097	0.0972	0.0010	0.0713	–
0.020	0.1838	0.0198	0.1503	0.0133	0.1190	–
0.050	0.2906	0.0498	0.2550	0.0454	0.2181	0.0323
0.100	0.4064	0.0998	0.3698	0.0964	0.3299	0.0863
0.200	0.5624	0.1998	0.5250	0.1972	0.4830	0.1892
0.500	0.8478	0.4999	0.8098	0.4979	0.7658	0.4919
1.000	1.1363	0.9999	1.0980	0.9982	1.0531	0.9931

because both inspections are coincident with each other. If $T_0^* + 1/\theta \geq T_S^* + c_T/c_D$ then there does not exist for positive $\widehat{c}_R$, i.e., inspection overtime cannot be rather than periodic inspection. □

4.4 Finite Interval

We take up a random inspection policy for a finite interval. The optimum policies for preventive maintenance, inspection, and cumulative damage models were summarized for a finite interval [6], [7, p. 59]. Suppose that the unit is checked only at successive working times S_j $(j = 1, 2, \ldots)$ in Fig. 4.1 for a specified interval $[0, S]$ $(0 < S < \infty)$. When S is a random variable, optimum policies will be discussed in Sect. 8.1.1.

We consider the following three cases: The probability that the unit fails and its failure is detected at random checking times is

$$\int_0^S \left\{ \sum_{j=0}^\infty \int_0^t [G(S-x) - G(t-x)] \, dG^{(j)}(x) \right\} dF(t),$$

the probability that it fails, however, its failure is not detected before time S is

$$\int_0^S \left[\sum_{j=0}^\infty \int_0^t \overline{G}(S-x) \, dG^{(j)}(x) \right] dF(t),$$

and probability that it does not fail during $[0, S]$ is $\overline{F}(S)$. Thus, the total expected cost during $[0, S]$ is

$C_S(G)$

$$= \int_0^S \left(\sum_{j=0}^{\infty} \int_0^t \left\{ \int_{t-x}^{S-x} [(j+1)c_R + c_D(x+y-t)] \, dG(y) \right\} dG^{(j)}(x) \right) dF(t)$$

$$+ \int_0^S \left\{ \sum_{j=0}^{\infty} \int_0^t [jc_R + c_D(S-t)]\overline{G}(S-x) \, dG^{(j)}(x) \right\} dF(t)$$

$$+ c_R \overline{F}(S) \sum_{j=0}^{\infty} j[G^{(j)}(S) - G^{(j+1)}(S)]$$

$$= c_R \int_0^S [1 - F(t)G(S-t)] \, dM(t) + c_D \left(\int_0^S F(t)\overline{G}(t) \, dt \right)$$

$$+ \int_0^S \left\{ \int_0^{S-x} [F(t+x) - F(x)]\overline{G}(t) \, dt \right\} dM(x) \Bigg), \qquad (4.66)$$

where c_R and c_D are given in (4.3) (Problem 7 in Sect. 4.5). Clearly, $C_S(G)$ agrees with (4.4) as $S \to \infty$.

In particular, when $G(t) = 1 - e^{-\theta t}$ and $F(t) = 1 - e^{-\lambda t}$ for $\theta > \lambda$, the total expected cost in (4.66) is (Problem 8 in Sect. 4.5)

$$C_S(\theta) = \frac{c_R \theta}{\lambda}(1 - e^{-\lambda S}) + (c_R \theta + c_D)\left(\frac{1 - e^{-\theta S}}{\theta} - \frac{e^{-\lambda S} - e^{-\theta S}}{\theta - \lambda} \right), \qquad (4.67)$$

which agrees with $C_R(\theta)$ in (4.21) as $S \to \infty$. Clearly,

$$\lim_{\theta \to \lambda} C_S(\theta) = c_R(1 - e^{-\lambda S}) + \frac{c_R \lambda + c_D}{\lambda}[1 - (1 + \lambda S)e^{-\lambda S}],$$

$$\lim_{\theta \to \infty} C_S(\theta) = \infty.$$

Thus, there exists a finite θ^* ($\lambda \le \theta^* < \infty$) which minimizes $C_S(\theta)$ (Problem 9 in Sect. 4.5).

4.5 Problems

1 Derive (4.4).
2 Derive (4.5).
3 Make that $T_S^*/(1/\theta^*) < \sqrt{2}$ in Table 4.3.
4 Prove that T_F^* increases with θ from T_S^* to ∞.

*5 When the failure time has a general distribution $F(t)$, consider the modified inspection models where the unit is checked at time $\min\{T, Y_k\}$ $(k = 1, 2, \ldots)$ for inspection first and time $\max\{T, Y_k\}$ for inspection last.

6 Prove that $T_O^* < T_L^*$, and compare inspection overtime with inspection first and last numerically.

7 Derive (4.66).

8 Derive (4.67) from (4.66) and compute optimum $1/\theta^*$, which minimizes $C_S(\theta)$ in (4.67).

*9 Consider the inspection policy with Nth random checks for a finite interval $[0, S]$.

References

1. Barlow RE, Proschan F (1965) Mathematical theory of reliability. Wiley, New York
2. Nakagawa T (2005) Maintenance theory of reliability. Springer, London
3. Blischke WR, Murthy DNP (2003) Case studies in reliability and maintenance. Wiley, New York
4. Christer AH (2002) A review of delay time analysis for modelling plant maintenance. In: Osaki S (ed) Stochastic models in reliability and maintenance. Springer, Berlin, pp 89–123
5. Wang W (2008) Delay time modelling. In: Kobacy KAH, Murthy DNP (eds) Complex system maintenance handbook. Springer, London, pp 345–370
6. Nakagawa T, Mizutani S (2009) A summary of maintenance policies for a finite interval. Reliab Eng Syst Saf 94:89–96
7. Nakagawa T (2008) Advanced reliability models and maintenance policies. Springer, London
8. Nakagawa T, Mizutani S, Chen M (2010) A summary of periodic and random inspection policies. Reliab Eng Syst Saf 95:906–911
9. Nakagawa T, Zhao X, Yun WY (2011) Optimal age replacement and inspection policies with random failure and replacement times. Inter J Reliab Qual Saf Eng 18:1–12
10. Nakagawa T (2011) Stochastic processes with applications to reliability theory. Springer, London

Chapter 5
Random Backup Policies

The random inspection policies of deriving how to detect failures have been summarized in Chap. 4. On the other hand, when failures are detected in the recovery technique for a database system, we have to execute the backup operation to the latest checkpoint [1, 2] and reconstruct the consistency of the database. It has been assumed in such models that any failures are always detected immediately, however, there is a loss time or cost, which might depend on the lapsed time for the backup operation between failure detection and the latest checking time. Optimum periodic and sequential checking times for such backup operation were derived analytically [3] and summarized [4, p. 123]. Furthermore, several backup policies for database systems with random working times were discussed [5–7], by applying the inspection policy to the backup policy.

This chapter summarizes such backup policies with random working times [5–7] and add new results: The total expected costs until the backup operation are obtained, and using these results, optimum periodic and random checking times are compared numerically in Sect. 5.1. It is shown that periodic policy is better than random one. In addition, when the checking cost for random policy is lower than that for periodic one, the random checking cost is computed numerically when both expected costs of periodic and random policies are the same.

In Sect. 5.2, when the system is checked at the Nth completion of working times, the expected cost is obtained and an optimum policy which minimizes it is derived when failure and working times are exponential. Furthermore, when a failure occurs, we execute the backup operation until the latest checking time and repeat such processes until the next checking time. An optimum policy which minimizes the expected cost per one work is discussed analytically and numerically.

In Sect. 5.3, as one of modified examples of backup policies, we consider the case where the failure is detected only at checking times and the backup operation is executed until the latest checking time. Two optimum backup policies which minimize the expected costs are derived.

In Sect. 5.4, we consider the following random checkpoint models. Most computer systems in offices and industries execute successively works each of which

© Springer-Verlag London 2014
T. Nakagawa, *Random Maintenance Policies*,
Springer Series in Reliability Engineering, DOI 10.1007/978-1-4471-6575-0_5

has a random processing time. In such systems, some failures often occur due to noises, human errors, and hardware faults. To detect and mask failures, some useful fault-tolerant computing techniques have been adopted [8, 9]. The simplest scheme in recovery techniques of failure detection is as follows [10]: We execute two independent modules, which compare two states at checkpoint times. If two states of each module do not match with each other, we go back to the newest checkpoint and make their retrials.

Several studies of deciding optimum checkpoint frequencies have been made. The performance and reliability of a double modular system with one spare module were evaluated [11, 12]. Furthermore, the performance of checkpoint schemes with task duplication was evaluated [13, 14]. The optimum instruction-retry period which minimizes the probability of the dynamic failure by a triple modular controller was derived [15]. Evaluation models with finite checkpoints and bounded rollback were discussed [16]. We introduce two types of checkpoints such as compare-checkpoint and compare-and-store-checkpoint, and using them, we consider three checkpoint schemes. Three schemes are compared and the best scheme among them is determined numerically. As one of examples, when a job has four works, six types of schemes are given and compared.

We use the following same notations denoted in Chap. 4: The system has a failure distribution $F(t)$ with finite mean μ $(0 < \mu < \infty)$, a failure density function $f(t)$, i.e., $F(t) \equiv \int_0^t f(u)\mathrm{d}u$, and $\overline{\Phi}(t) \equiv 1 - \Phi(t)$ for any function $\Phi(t)$. The random times such as working and processing times for a job are denoted by Y_j $(j = 1, 2, \ldots)$, where Y_j is independent and has an identical distribution $G(t)$ with finite mean $1/\theta$ $(0 < \theta < \infty)$, and $S_j \equiv \sum_{i=1}^{j} Y_i$ $(j = 1, 2, \ldots)$ and $S_0 \equiv 0$. Then, the probability that the system works exactly j times in $[0, t]$ is $\Pr\{S_j < t \leq S_{j+1}\} = G^{(j)}(t) - G^{(j+1)}(t)$, where $G^{(j)}(t)$ $(j = 1, 2, \ldots)$ denote the j-fold Stieltjes convolution of $G(t)$ with itself and $\Phi^{(0)}(t) \equiv 1$ for $t \geq 0$. In addition, $M(t) \equiv \sum_{j=1}^{\infty} G^{(j)}(t)$, which represents the expected number of random working times in $[0, t]$, and it is called a renewal function in stochastic processes [17, p. 50].

5.1 Periodic and Random Backup Times

We consider the process of backup policies for a job with random working times: Suppose that the system is checked at successive random times S_j $(j = 1, 2, \ldots)$ such as working and processing times, and also at periodic times kT $(k = 1, 2, \ldots)$ for a specified $T > 0$ in Fig. 4.1. When failures occur, they are detected immediately, and the backup operation is executed until the latest checking time to restore the system consistency. The process ends at the backup operation and starts newly from this checking point.

We introduce the following costs or overheads for the above process: c_T and c_R are the respective costs for periodic and random checks. When a failure occurs at time t between kT and $(k + 1)T$ or S_{j+1}, the backup operation is executed from

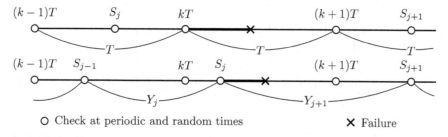

O Check at periodic and random times   ✕ Failure

Fig. 5.1 Process of periodic and random backup policies

the failure time t to the latest checking time kT in Fig. 5.1. This incurs a loss cost $c_D(t - kT)$ which includes all costs resulting from the working time from kT to t and the backup operation from t to kT. On the other hand, when a failure occurs at time t between S_j and $(k + 1)T$ or S_{j+1}, this incurs a loss cost $c_D(t - S_j)$.

The probability that the process goes back to periodic check due to some failure is

$$\sum_{k=0}^{\infty} \int_{kT}^{(k+1)T} \left[\sum_{j=0}^{\infty} \int_0^{kT} \overline{G}(t - x)\, dG^{(j)}(x) \right] dF(t), \tag{5.1}$$

and the probability that it goes back to random check is

$$\sum_{k=0}^{\infty} \int_{kT}^{(k+1)T} \left[\sum_{j=0}^{\infty} \int_{kT}^{t} \overline{G}(t - x)\, dG^{(j)}(x) \right] dF(t), \tag{5.2}$$

where $(5.1) + (5.2) = 1$ (Problem 1 in Sect. 5.6). Therefore, the total expected cost until the backup operation is

$$C(T) = \sum_{k=0}^{\infty} \int_{kT}^{(k+1)T} \left\{ \sum_{j=0}^{\infty} \int_0^{kT} [c_T k + c_R j + c_D(t - kT)] \overline{G}(t - x)\, dG^{(j)}(x) \right\} dF(t)$$

$$+ \sum_{k=0}^{\infty} \int_{kT}^{(k+1)T} \left\{ \sum_{j=0}^{\infty} \int_{kT}^{t} [c_T k + c_R j + c_D(t - x)] \overline{G}(t - x)\, dG^{(j)}(x) \right\} dF(t)$$

$$= c_T \sum_{k=1}^{\infty} \overline{F}(kT) + c_R \int_0^{\infty} M(t)\, dF(t) + c_D \mu$$

$$- c_{\mathrm{D}} \left\{ \sum_{k=0}^{\infty} (kT) \int_{kT}^{(k+1)T} \left[\sum_{j=0}^{\infty} \int_{0}^{kT} \overline{G}(t-x)\,\mathrm{d}G^{(j)}(x) \right] \mathrm{d}F(t) \right.$$

$$\left. + \sum_{k=0}^{\infty} \int_{kT}^{(k+1)T} \left[\sum_{j=0}^{\infty} \int_{kT}^{t} x\overline{G}(t-x)\,\mathrm{d}G^{(j)}(x) \right] \mathrm{d}F(t) \right\}. \tag{5.3}$$

In particular, when $Y_j \equiv \infty$, i.e., $G(t) \equiv 0$ for any $t \geq 0$, the system is checked only at periodic times kT $(k = 1, 2, \ldots)$, and the total expected cost in (5.3) is

$$C(T) = (c_T - c_{\mathrm{D}}T) \sum_{k=1}^{\infty} \overline{F}(kT) + c_{\mathrm{D}}\mu, \tag{5.4}$$

which agrees with (5.55) of [4, p. 95]. When $T = \infty$, the system is checked only at random times S_j $(j = 1, 2, \ldots)$, and the total expected cost is (Problem 2 in Sect. 5.6)

$$C(G) = c_{\mathrm{R}} \int_{0}^{\infty} M(t)\,\mathrm{d}F(t) + c_{\mathrm{D}}\mu - c_{\mathrm{D}} \int_{0}^{\infty} \left[\sum_{j=0}^{\infty} \int_{0}^{t} x\overline{G}(t-x)\,\mathrm{d}G^{(j)}(x) \right] \mathrm{d}F(t), \tag{5.5}$$

which agrees with [6].

We compare periodic and random policies when $F(t) = 1 - e^{-\lambda t}$ $(0 < \lambda < \infty)$ and $G(t) = 1 - e^{-\theta t}$ $(0 < \theta < \infty)$. In this case, the total expected cost in (5.4) is

$$C_{\mathrm{P}}(T) = \frac{c_T - c_{\mathrm{D}}T}{e^{\lambda T} - 1} + \frac{c_{\mathrm{D}}}{\lambda}. \tag{5.6}$$

Clearly,

$$C_{\mathrm{P}}(0) \equiv \lim_{T \to 0} C_{\mathrm{P}}(T) = \infty, \qquad C_{\mathrm{P}}(\infty) \equiv \lim_{T \to \infty} C_{\mathrm{P}}(T) = \frac{c_{\mathrm{D}}}{\lambda}.$$

Thus, there exists an optimum T^* $(0 < T^* \leq \infty)$ which minimizes (5.6). Differentiating $C_{\mathrm{P}}(T)$ with respect to T and setting it equals to zero,

$$\lambda T - (1 - e^{-\lambda T}) = \frac{c_T}{c_{\mathrm{D}}/\lambda}, \tag{5.7}$$

whose left-hand increases strictly from 0 to ∞. Thus, there exists a finite and unique T^* $(0 < T^* < \infty)$ which satisfies (5.7), and the resulting cost is

$$\frac{C_P(T^*)}{c_D/\lambda} = 1 - e^{-\lambda T^*}. \tag{5.8}$$

Similarly, the total expected cost in (5.5) is

$$C_R(\theta) = \frac{c_R \theta}{\lambda} + \frac{c_D}{\theta + \lambda}. \tag{5.9}$$

An optimum θ^* which minimizes $C_R(\theta)$ is easily given by

$$\frac{\lambda}{\theta + \lambda} = \sqrt{\frac{c_R \lambda}{c_D}}, \tag{5.10}$$

and the resulting cost is

$$\frac{C_R(\theta^*)}{c_D/\lambda} = \frac{\lambda}{\theta^* + \lambda} \left(\frac{\theta^*}{\theta^* + \lambda} + 1 \right). \tag{5.11}$$

If $\lambda c_R/c_D \geq 1$, then $1/\theta^* = \infty$, and $C_R(0) = C_P(\infty) = c_D/\lambda$.

Example 5.1 (*Checking time for exponential failure time*) Table 5.1 presents optimum T^*, $1/\theta^*$ and their costs $C_P(T^*)/c_D$, $C_R(\theta^*)/c_D$ for c_T/c_D when $c_T = c_R$ and $\lambda = 1$. This indicates that $T^* > 1/\theta^*$ when c_T/c_D is small, and $C_P(T^*) < C_R(\theta^*)$, i.e., periodic policy is better than random one, as shown in Sect. 4.1. From (5.7) and (5.10), when

$$T - \left(\frac{T}{1 + T} \right)^2 = 1 - e^{-T},$$

$T^* = 1/\theta^* = 0.694$. In this case,

$$\frac{c_T}{c_D} = \left(\frac{T^*}{1 + T^*} \right)^2 = 0.168.$$

That is, when $c_T/c_D = 0.168$, $T^* = 1/\theta^* = 0.694$, and $T^* > 1/\theta^*$ for $c_T/c_D < 0.168$ and $T^* < 1/\theta^*$ for $c_T/c_D > 0.168$. $\square$

It has been assumed that $c_T = c_R$ in Table 5.1. In general, a random checking cost c_R would be lower than a periodic one because the system is checked at random. We compute a random checking cost $\widehat{c}_R$ when both expected costs of two backup policies are the same. From Table 5.1, we compute $1/\widehat{\theta}$ for c_T/c_D when

$$\frac{C_P(T^*)}{c_D} = 1 - e^{-T^*} = \frac{2\widehat{\theta} + 1}{(\widehat{\theta} + 1)^2},$$

Table 5.1 Optimum T^*, $1/\theta^*$ and their cost rates when $c_T = c_R$ and $\lambda = 1$

c_T/c_D	T^*	$C_P(T^*)/c_D$	$1/\theta^*$	$C_R(\theta^*)/c_D$
0.001	0.045	0.044	0.033	0.063
0.002	0.064	0.062	0.047	0.088
0.005	0.102	0.097	0.076	0.136
0.010	0.145	0.135	0.111	0.190
0.020	0.207	0.187	0.165	0.263
0.050	0.334	0.284	0.288	0.397
0.100	0.483	0.383	0.463	0.532
0.200	0.707	0.507	0.809	0.694
0.500	1.198	0.698	2.414	0.914
1.000	1.841	0.841	∞	1.000

Table 5.2 Values of $1/\widehat{\theta}$, $\widehat{c}_R/c_D$ and $\widehat{c}_R/c_T$ when $\lambda = 1$

c_T/c_D	$1/\widehat{\theta}$	$\widehat{c}_R/c_D$	$\widehat{c}_R/c_T$
0.001	0.023	0.0005	0.5000
0.002	0.032	0.0010	0.5000
0.005	0.052	0.0025	0.5000
0.010	0.075	0.0049	0.4900
0.020	0.109	0.0097	0.4850
0.050	0.182	0.0237	0.4740
0.100	0.273	0.0460	0.4600
0.200	0.424	0.0887	0.4435
0.500	0.820	0.2030	0.4060
1.000	1.511	0.3623	0.3623

which decreases strictly with $\widehat{\theta}$ from 1 to 0, and compute

$$\frac{\widehat{c}_R}{c_D} = \left(\frac{1}{\widehat{\theta}+1}\right)^2 .$$

Example 5.2 (Random cost for exponential failure time) Table 5.2 presents $1/\widehat{\theta}$, $\widehat{c}_R/c_D$ and $\widehat{c}_R/c_T$, and indicates that $\widehat{c}_R$ is a little lower than the half of c_T. In other words, when the random checking cost is the half of the periodic one, both expected costs are almost the same (Problem 3 in Sect. 5.6). For example, $C_P(T^*)/c_T$ when $c_T/c_D = 0.002, 0.010, 0.020, 0.100, 0.200, 1.000$, are almost equal to $C_R(\theta^*)/c_T$ when $c_T/c_D = 0.001, 0.005, 0.010, 0.050, 0.100, 0.500$, respectively, in Table 5.1. As c_T/c_D becomes larger, T^* becomes larger, i.e., $\widehat{\theta}$ becomes 0, and hence, $\widehat{c}_R/c_D$ becomes 1 and $\widehat{c}_R/c_T$ approaches to 0. □

5.2 Optimum Checking Time

When $G(t) = 1 - e^{-\theta t}$, the total expected cost in (5.3) is

$$C_P(T) = c_T \sum_{k=1}^{\infty} \overline{F}(kT) + c_R \theta \mu + \frac{c_D}{\theta} \sum_{k=0}^{\infty} \int_{kT}^{(k+1)T} [1 - e^{-\theta(t-kT)}] \, dF(t). \quad (5.12)$$

In particular, when $F(t) = 1 - e^{-\lambda t}$,

$$C_P(T) = \frac{c_T}{e^{\lambda T} - 1} + \frac{c_R \theta}{\lambda} + \frac{c_D}{\theta} \left[1 - \frac{\lambda}{\theta + \lambda} \frac{1 - e^{-(\theta + \lambda)T}}{1 - e^{-\lambda T}} \right]. \quad (5.13)$$

Clearly,

$$C_P(0) \equiv \lim_{T \to 0} C_P(T) = \infty,$$

$$C_P(\infty) \equiv \lim_{T \to \infty} C_P(T) = \frac{c_R \theta}{\lambda} + \frac{c_D}{\theta + \lambda} = C_R(\theta),$$

which agrees with (5.9). We find an optimum T_P^* ($0 < T_P^* \le \infty$) which minimizes $C_P(T)$ in (5.13). Differentiating $C_P(T)$ with respect to T and setting it equal to zero,

$$\frac{1 - e^{-\theta T}}{\theta} - \frac{1 - e^{-(\theta + \lambda)T}}{\theta + \lambda} = \frac{c_T}{c_D}, \quad (5.14)$$

whose left-hand increases strictly with T from 0 to $\lambda/[\theta(\theta + \lambda)]$. Therefore, if $\lambda/(\theta + \lambda) > c_T/(c_D/\theta)$ then there exists a finite and unique T_P^* ($0 < T_P^* < \infty$) which satisfies (5.14). In addition, the left-hand side of (5.14) increases strictly with $1/\theta$, i.e., T_P^* decreases with $1/\theta$ to T^* given in (5.7) (Problem 4 in Sect. 5.6).

Next, the system is checked at random times S_j and also at successive times T_k ($k = 1, 2, \ldots$), where $T_0 \equiv 0$. Then, replacing kT in (5.3) with T_k formally,

$$C_P(T_1, T_2, \ldots) = c_T \sum_{k=1}^{\infty} \overline{F}(T_k) + c_R \int_0^{\infty} M(t) \, dF(t) + c_D \mu$$

$$- c_D \left\{ \sum_{k=0}^{\infty} T_k \int_{T_k}^{T_{k+1}} \left[\sum_{j=0}^{\infty} \int_0^{T_k} \overline{G}(t - x) \, dG^{(j)}(x) \right] dF(t) \right.$$

$$\left. + \sum_{k=0}^{\infty} \int_{T_k}^{T_{k+1}} \left[\sum_{j=0}^{\infty} \int_{T_k}^{t} x \overline{G}(t - x) \, dG^{(j)}(x) \right] dF(t) \right\}. \quad (5.15)$$

In particular, when $G(t) = 1 - e^{-\theta t}$, from (5.12),

$$C_P(T_1, T_2, \ldots) = c_T \sum_{k=1}^{\infty} \overline{F}(T_k) + c_R \theta \mu + \frac{c_D}{\theta} \sum_{k=0}^{\infty} \int_{T_k}^{T_{k+1}} [1 - e^{-\theta(t-T_k)}] \, dF(t).$$

(5.16)

Differentiating $C_P(T_1, T_2, \ldots)$ with respect to T_k and setting it equal to zero,

$$\frac{1}{f(T_k)} \int_{T_k}^{T_{k+1}} e^{-\theta(t-T_k)} dF(t) = \frac{1}{\theta} [1 - e^{-\theta(T_k - T_{k-1})}] - \frac{c_T}{c_D} \quad (k = 1, 2, \ldots). \quad (5.17)$$

When the system is checked only at successive times S_j, i.e., $\theta \to 0$, (5.17) becomes

$$\frac{F(T_{k+1}) - F(T_k)}{f(T_k)} = T_k - T_{k-1} - \frac{c_T}{c_D},$$

(5.18)

which agrees with (5.54) of [4, p. 95]. Therefore, by using Algorithm [1, p. 112], we can compute an optimum schedule which satisfies (5.18).

Example 5.3 (*Checking time for Weibull failure time*) Table 5.3 presents optimum T_k^* ($k = 1, 2, \ldots$) for $1/\theta = 0.1, 0.5, \infty$ when $F(t) = 1 - \exp(-t^2)$ and $c_T/c_D = 0.02$. This indicates that T_k^* decreases with $1/\theta$, however, varies a little for $1/\theta$, and increases gradually with k. This has a similar tendency to Table 4.2., i.e., if the mean random time $1/\theta$ becomes small, T_k^* becomes large because the system is sometimes checked at random times. ☐

Table 5.3 Optimum T_k^* when $F(t) = 1 - e^{-t^2}$ and $c_T/c_D = 0.02$

k	$1/\theta = 0.1$	$1/\theta = 0.5$	$1/\theta = \infty$
1	1.344	0.339	0.290
2	1.629	0.570	0.505
3	1.862	0.766	0.688
4	2.066	0.940	0.853
5	2.251	1.100	1.006
6	2.421	1.249	1.150
7	2.577	1.390	1.289
8	2.717	1.524	1.423
9	2.838	1.652	1.554
10	2.934	1.776	1.686

5.3 Random Backup

5.3.1 N Works

Suppose that the system is checked at every Nth ($N = 1, 2, \ldots$) random times, i.e., $S_N, S_{2N}, \ldots$. By replacing $G(t)$ with $G^{(N)}(t)$ in (5.5) formally, the total expected cost until backup operation is [6]

$$
C_{R1}(N) = c_R \int_0^\infty M^{(N)}(t) \, dF(t) + c_D \mu
$$

$$
- c_D \int_0^\infty \left\{ \sum_{j=0}^\infty \int_0^t x \left[1 - G^{(N)}(t - x) \right] dG^{(jN)}(x) \right\} dF(t)
$$

$$
(N = 1, 2, \ldots), \qquad (5.19)
$$

where $M^{(N)}(t) \equiv \sum_{j=1}^\infty G^{(jN)}(t)$. In particular, when $F(t) = 1 - e^{-\lambda t}$ and $G(t) = 1 - e^{-\theta t}$ (Problem 5 in Sect. 5.6),

$$
C_{R1}(N) = c_R \frac{A^N}{1 - A^N} + \frac{c_D}{\lambda} \frac{(1 - A)^2}{1 - A^N} \sum_{j=1}^N j A^{j-1}, \qquad (5.20)
$$

where $A \equiv \theta/(\theta + \lambda) = \tilde{R}(\theta)$ in Sect. 1.2. From the inequality $C_{R1}(N + 1) - C_{R1}(N) \geq 0$,

$$
(1 - A) \sum_{j=1}^N (1 - A^j) \geq \frac{c_R}{c_D/\lambda}. \qquad (5.21)
$$

Therefore, there exists a finite and unique minimum N_1^* ($1 \leq N_1^* < \infty$) which satisfies (5.21). Note that N_1^* increases with A, i.e., N_1^* decreases with $1/\theta$ from ∞ to a minimum integer such that $N \geq c_R/(c_D/\lambda)$.

Next, we obtain the total expected cost until N works have been completed [7]: First, suppose that $N = 1$. When the system fails between S_j and S_{j+1}, we carry out the backup operation to the latest checking time S_j and reexecute the work again. It is assumed that the system becomes like new by the backup operation. Then, the expected cost until the completion of one work is given by the renewal equation

$$
\tilde{C}_R(1) = c_R \int_0^\infty \overline{F}(t) \, dG(t) + \int_0^\infty [c_D t + \tilde{C}_R(1)] \overline{G}(t) \, dF(t).
$$

Solving this renewal equation for $\widetilde{C}_R(1)$,

$$\widetilde{C}_R(1) = c_R + \frac{c_D \int_0^\infty t\overline{G}(t)\, dF(t)}{\int_0^\infty G(t)\, dF(t)}. \tag{5.22}$$

By replacing $G(t)$ with $G^{(N)}(t)$ formally, the expected cost until the completion of N works is

$$\widetilde{C}_R(N) = c_R + \frac{c_D \int_0^\infty t[1 - G^{(N)}(t)]\, dF(t)}{\int_0^\infty G^{(N)}(t)\, dF(t)} \qquad (N = 1, 2, \ldots). \tag{5.23}$$

As one of appropriate objective functions, we adopt the expected cost per one work given by

$$C_{R2}(N) \equiv \frac{\widetilde{C}_R(N)}{N} = \frac{1}{N}\left\{ c_R + \frac{c_D \int_0^\infty t[1 - G^{(N)}(t)]\, dF(t)}{\int_0^\infty G^{(N)}(t)\, dF(t)} \right\} \qquad (N = 1, 2, \ldots). \tag{5.24}$$

In particular, when $F(t) = 1 - e^{-\lambda t}$ and $G(t) = 1 - e^{-\theta t}$,

$$C_{R2}(N) = \frac{1}{N}\left[c_R + \frac{c_D}{\lambda}\frac{(1 - A)^2}{A^{N+1}} \sum_{j=1}^{N} j A^j \right]. \tag{5.25}$$

From the inequality $C_{R2}(N + 1) - C_{R2}(N) \geq 0$,

$$\frac{1 - A}{A^{N+1}} \sum_{j=1}^{N} (1 - A^j) \geq \frac{c_R}{c_D/\lambda}, \tag{5.26}$$

whose left-hand side increases strictly from $[(1 - A)/A]^2 = (\lambda/\theta)^2$ to ∞. Therefore, there exists a finite and unique N_2^* ($1 \leq N_2^* < \infty$) which satisfies (5.26). If $(\lambda/\theta)^2 \geq c_R/(c_D/\lambda)$, then $N_2^* = 1$, i.e., we should place checking times at every completion of works. It can be seen that N_2^* decreases with $1/\theta$ from ∞ to 1 (Problem 6 in Sect. 5.6). Compared (5.21) with (5.26), $N_1^* \geq N_2^*$.

Example 5.4 (*Checking number for exponential failure time*) Table 5.4 presents optimum N_1^* and N_2^* for $1/\theta$ when $c_R/c_D = 0.1$ and $\lambda = 1$. This indicates that both N_1^* and N_2^* decrease with $1/\theta$ to 1 and $N_1^* \geq N_2^*$. If $1/\theta \geq \sqrt{0.1} \approx 0.316$ then $N_2^* = 1$, and if $(1 - A)^2 \geq 0.1$, i.e., $1/\theta \geq (1 + \sqrt{10})/9 \approx 0.462$ then $N_1^* = 1$. Compared to Table 5.1, $N_1^*/\theta \approx 1/\theta^*$ when $c_T/c_D = c_R/c_D = 0.1$. $\qquad\square$

Furthermore, we derive the mean time to the completion of one work, which is given by the renewal function

Table 5.4 Optimum N_1^* and N_2^* when $c_R/c_D = 0.1$ and $\lambda = 1$

$1/\theta$	N_1^*	N_2^*
0.01	49	39
0.02	25	20
0.03	16	13
0.04	12	10
0.05	10	8
0.06	8	7
0.07	7	6
0.08	6	5
0.09	6	5
0.10	5	4
0.20	3	2
0.50	1	1

$$L_R(1) = \int_0^\infty t \overline{F}(t) \, dG(t) + \int_0^\infty [t + L_R(1)] \overline{G}(t) \, dF(t),$$

i.e.,

$$L_R(1) = \frac{\int_0^\infty \overline{G}(t) \overline{F}(t) \, dt}{\int_0^\infty G(t) \, dF(t)}. \tag{5.27}$$

Thus, the mean time to the completion of N works is, replacing $G(t)$ with $G^{(N)}(t)$,

$$L_R(N) = \frac{\int_0^\infty [1 - G^{(N)}(t)] \overline{F}(t) \, dt}{\int_0^\infty G^{(N)}(t) \, dF(t)}. \tag{5.28}$$

Therefore, the expected cost rate for N works is, from (5.23) to (5.28),

$$C_{R3}(N) \equiv \frac{\widetilde{C}_R(N)}{L_R(N)} = \frac{c_R \int_0^\infty G^{(N)}(t) \, dF(t) + c_D \int_0^\infty t[1 - G^{(N)}(t)] \, dF(t)}{\int_0^\infty [1 - G^{(N)}(t)] \overline{F}(t) \, dt}$$

$$(N = 1, 2, \ldots). \tag{5.29}$$

In particular, when $F(t) = 1 - e^{-\lambda t}$ and $G(t) = 1 - e^{-\theta t}$,

$$\frac{C_{R3}(N)}{\lambda} = \frac{c_R A^N + (c_D/\lambda)(1 - A)^2 \sum_{j=1}^N j A^{j-1}}{1 - A^N}, \tag{5.30}$$

which agrees with $C_{R1}(N)$ in (5.20). In this case, an optimum policy which minimizes $C_{R3}(N)$ corresponds to that for $C_{R1}(N)$ (Problem 7 in Sect. 5.6).

5.3.2 6 Works

We consider the optimization problem in which how we should plan the checking schedule of six random works in Fig. 5.2 [7]: The checking times are placed at (i) $n = 6$, (ii) $n = 3, 6$, (iii) $n = 2, 4, 6$, and (iv) $n = 1, 2, 3, 4, 5, 6$ for Model i ($i = 1, 2, 3, 4$), respectively.

By the similar method of obtaining (5.23), the total expected cost $C_R(i)$ of Model i ($i = 1, 2, 3, 4$) until the completion of six works is (Problem 8 in Sect. 5.6)

$$C_R(1) = c_R + \frac{c_D \int_0^\infty t[1 - G^{(6)}(t)]\,dF(t)}{\int_0^\infty G^{(6)}(t)\,dF(t)},$$

$$C_R(2) = 2\left[c_R + \frac{c_D \int_0^\infty t[1 - G^{(3)}(t)]\,dF(t)}{\int_0^\infty G^{(3)}(t)\,dF(t)}\right],$$

$$C_R(3) = 3\left[c_R + \frac{c_D \int_0^\infty t[1 - G^{(2)}(t)]\,dF(t)}{\int_0^\infty G^{(2)}(t)\,dF(t)}\right],$$

$$C_R(4) = 6\left[c_R + \frac{c_D \int_0^\infty t\overline{G}(t)\,dF(t)}{\int_0^\infty \overline{G}(t)\,dF(t)}\right].$$

In particular, when $F(t) = 1 - e^{-\lambda t}$ and $G(t) = 1 - e^{-\theta t}$, the expected costs are

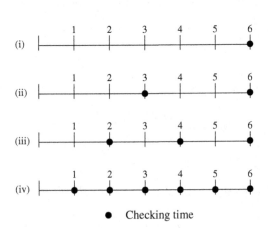

Fig. 5.2 Process of checking times for six works

● Checking time

$$C_R(1) = c_R + \frac{c_D}{\theta} \frac{1-A}{A^5} \sum_{j=0}^{5} (j+1)A^j,$$

$$C_R(2) = 2\left[c_R + \frac{c_D}{\theta} \frac{1-A}{A^2} \sum_{j=0}^{2} (j+1)A^j \right],$$

$$C_R(3) = 3\left[c_R + \frac{c_D}{\theta} \frac{1-A}{A} \sum_{j=0}^{1} (j+1)A^j \right],$$

$$C_R(4) = 6\left[c_R + \frac{c_D}{\theta}(1-A) \right].$$

We can compare $C_R(i)$ with $C_R(i+1)$ ($i = 1, 2, 3$) as follows:

$$C_R(2) \le C_R(1) \Leftrightarrow \frac{(1-A^3)(1+A+A^2)}{A^5} \ge \frac{c_R}{c_D/\theta},$$

$$C_R(3) \le C_R(2) \Leftrightarrow \frac{(1-A)(2+A)}{A^2} \ge \frac{c_R}{c_D/\theta},$$

$$C_R(4) \le C_R(3) \Leftrightarrow \frac{1-A}{A} \ge \frac{c_R}{c_D/\theta}.$$

Example 5.5 (Checking time for six works) Table 5.5 presents $C_R(i)/(c_D/\theta)$ for $c_R/(c_D/\theta)$ when $A = 5/6$, i.e., $\lambda/\theta = 0.2$. This indicates that the best model with minimum cost moves to Model i ($i = 1, 2, 3, 4$) as $c_R/(c_D/\theta)$ becomes smaller from 0.7 to 0.08, and their costs increase with $c_R/(c_D/\theta)$. In other words, as $c_R/(c_D/\theta)$ decreases, we should place more checking points and the expected costs decrease. □

Table 5.5 Expected costs $C_R(i)/(c_D/\theta)$ of Model i ($i = 1, 2, 3, 4$) when $A = 5/6$

$\frac{c_R}{c_D/\theta}$	$\frac{C_R(1)}{c_D/\theta}$	$\frac{C_R(2)}{c_D/\theta}$	$\frac{C_R(3)}{c_D/\theta}$	$\frac{C_R(4)}{c_D/\theta}$
0.7	2.35[a]	2.44	2.91	4.75
0.6	2.25	2.24[a]	2.61	4.15
0.5	2.15	2.04[a]	2.31	3.55
0.4	2.05	1.84[a]	2.01	2.95
0.3	1.95	1.64	1.71	2.35
0.2	1.85	1.44	1.41[a]	1.75
0.1	1.75	1.24	1.11[a]	1.15
0.09	1.74	1.22	1.08[a]	1.09
0.08	1.73	1.20	1.05	1.03[a]

[a] Minimum value

5.4 Random Backup for Continuous Model

It has been assumed until now that the failure is detected immediately. However, consider a continuous production system [18]. Failures of such systems are detected only at checking times. Suppose that the failure is detected only at random checking times and the backup operation is executed until the latest checking time in Fig. 5.3. This incurs a loss cost from the latest checking time S_j to failure detection.

By the similar method of obtaining (5.3), the total expected cost until backup operation is (Problem 9 in Sect. 5.6)

$$
C_C(1) = \sum_{j=0}^{\infty} \int_0^{\infty} \left(\int_0^y \left\{ \int_{y-t}^{\infty} [c_R(j+1) + c_D x] \, dG(x) \right\} dG^{(j)}(t) \right) dF(y)
$$

$$
= c_D \sum_{j=0}^{\infty} \int_0^{\infty} \left\{ \int_0^{\infty} x[F(t+x) - F(t)] \, dG(x) \right\} dG^{(j)}(t)
$$

$$
+ c_R \sum_{j=0}^{\infty} \int_0^{\infty} G^{(j)}(t) \, dF(t). \tag{5.31}
$$

Furthermore, suppose that the system is checked at every Nth ($N = 1, 2, \ldots$) random times, $S_N, S_{2N}, \ldots$ denoted in Sect. 5.3. By replacing $G(t)$ with $G^{(N)}(t)$ in (5.31), the total expected cost until backup operation is

$$
C_{C1}(N) = c_D \sum_{j=0}^{\infty} \int_0^{\infty} \left\{ \int_0^{\infty} x[F(t+x) - F(t)] \, dG^{(N)}(x) \right\} dG^{(jN)}(t)
$$

$$
+ c_R \sum_{j=0}^{\infty} \int_0^{\infty} G^{(jN)}(t) \, dF(t) \quad (N = 1, 2, \ldots). \tag{5.32}
$$

In particular, when $F(t) = 1 - e^{-\lambda t}$ and $G(t) = 1 - e^{-\theta t}$,

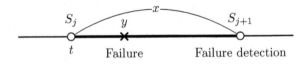

Fig. 5.3 Backup policy for a continuous model

$$C_{C1}(N) = c_R \sum_{j=0}^{\infty} (A^N)^j + c_D N \left(\frac{1}{\theta} - \frac{A^N}{\theta + \lambda} \right) \sum_{j=0}^{\infty} (A^N)^j$$

$$= \frac{c_R}{1 - A^N} + \frac{c_D}{\lambda} \frac{N(1 - A)(1 - A^{N+1})}{A(1 - A^N)}, \tag{5.33}$$

where $A \equiv \theta/(\theta + \lambda)$. We find an optimum N_3^* which minimizes $C_{C1}(N)$. From the inequality $C_{C1}(N + 1) - C_{C1}(N) \geq 0$,

$$\frac{(1 - A)^2}{A} \left[\frac{(1 - A^N)(1 - A^{N+2})}{(1 - A)^2 A^N} - N \right] \geq \frac{c_R}{c_D/\lambda}, \tag{5.34}$$

whose bracket on the left-hand increases strictly with N to ∞ (Problem 10 in Sect. 5.6). Therefore, there exists a finite and unique minimum N_3^* $(1 \leq N_3^* < \infty)$ which satisfies (5.34).

Next, when the system fails between S_j and S_{j+1}, we carry out the backup operation to S_j and reexecute the work again. By the similar method of obtaining (5.22), the expected cost until the completion of one work is given by the renewal equation

$$\tilde{C}_C(1) = c_R \int_0^{\infty} \overline{F}(t) \, dG(t) + \int_0^{\infty} [c_D t + \tilde{C}_C(1)] F(t) \, dG(t),$$

i.e.,

$$\tilde{C}_C(1) = c_R + \frac{c_D \int_0^{\infty} t F(t) \, dG(t)}{\int_0^{\infty} \overline{F}(t) \, dG(t)}. \tag{5.35}$$

By replacing $G(t)$ with $G^{(N)}(t)$ formally, the expected cost until the completion of N works is

$$\tilde{C}_C(N) = c_R + \frac{c_D \int_0^{\infty} t F(t) \, dG^{(N)}(t)}{\int_0^{\infty} \overline{F}(t) \, dG^{(N)}(t)} \qquad (N = 1, 2, \ldots). \tag{5.36}$$

Thus, the expected cost per one work is

$$C_{C2}(N) \equiv \frac{\tilde{C}_C(N)}{N} = \frac{c_R}{N} + \frac{c_D}{N} \frac{\int_0^{\infty} t F(t) \, dG^{(N)}(t)}{\int_0^{\infty} \overline{F}(t) \, dG^{(N)}(t)} \qquad (N = 1, 2, \ldots). \tag{5.37}$$

In particular, when $F(t) = 1 - e^{-\lambda t}$ and $G(t) = 1 - e^{-\theta t}$,

$$C_{C2}(N) = \frac{c_R}{N} + \frac{c_D}{\lambda} \frac{(1 - A)(1 - A^{N+1})}{A^{N+1}}. \tag{5.38}$$

From the inequality $C_{C2}(N+1) - C_{C2}(N) \geq 0$,

$$\left(\frac{1-A}{A}\right)^2 \frac{N(N+1)}{A^N} \geq \frac{c_R}{c_D/\lambda}, \tag{5.39}$$

whose left-hand increases strictly to ∞. Therefore, there exists a finite and unique N_4^* ($1 \leq N_4^* < \infty$) which satisfies (5.39). The left-hand side decreases with A, i.e., N_4^* decreases with $1/\theta$ from ∞ to 1, and $N_4^* \leq N_3^*$ (Problem 11 in Sect. 5.6).

Furthermore, the mean time to the completion of one work is given by the renewal function

$$L_C(1) = \int_0^\infty t\overline{F}(t)\,dG(t) + \int_0^\infty [t + L_C(1)]F(t)\,dG(t),$$

i.e.,

$$L_C(1) = \frac{1}{\theta \int_0^\infty \overline{F}(t)\,dG(t)}. \tag{5.40}$$

Thus, the mean time to the completion of N works is, replacing $G(t)$ and $1/\theta$ in (5.40) with $G^{(N)}(t)$ and N/θ, respectively,

$$L_C(N) = \frac{N}{\theta \int_0^\infty \overline{F}(t)\,dG^{(N)}(t)}. \tag{5.41}$$

Therefore, the expected cost rate for N works is, from (5.36) and (5.41),

$$C_{C3}(N) \equiv \frac{\widetilde{C}_C(N)}{L_C(N)} = \frac{\theta}{N}\left[c_R\int_0^\infty \overline{F}(t)\,dG^{(N)}(t) + c_D\int_0^\infty tF(t)\,dG^{(N)}(t)\right]$$
$$(N = 1, 2, \ldots). \tag{5.42}$$

In particular, when $F(t) = 1 - e^{-\lambda t}$ and $G(t) = 1 - e^{-\theta t}$,

$$\frac{C_{C3}(N)}{\lambda} = c_R\frac{A^{N+1}}{N(1-A)} + \frac{c_D}{\lambda}(1 - A^{N+1}). \tag{5.43}$$

From the inequality $C_{C3}(N+1) - C_{C3}(N) \geq 0$,

$$\frac{N(N+1)(1-A)^2}{N(1-A)+1} \geq \frac{c_R}{c_D/\lambda}, \tag{5.44}$$

Table 5.6 Optimum N_3^*, N_4^* and N_5^* when $c_R/c_D = 0.1$ and $\lambda = 1$

$1/\theta$	N_3^*	N_4^*	N_5^*
0.01	32	28	37
0.02	16	14	19
0.03	11	9	13
0.04	8	7	10
0.05	6	6	8
0.06	5	5	6
0.07	5	4	6
0.08	4	4	5
0.09	4	4	4
0.10	3	3	4
0.20	2	2	2
0.50	1	1	1

whose left-hand side increases strictly with N to ∞. Therefore, there exists a finite and unique minimum N_5^* ($1 \le N_5^* < \infty$) which satisfies (5.44), and decreases with $1/\theta$ because the left-hand side decreases with A. Compared (5.44) with (5.39), $N_5^* \ge N_4^*$, and compared (5.44) with (5.34), $N_5^* \ge N_3^*$, i.e., $N_5^* \ge N_3^* \ge N_4^*$ (Problem 12 in Sect. 5.6).

Example 5.6 (*Checking number for continuous model*) Table 5.6 presents optimum N_3^* and N_4^* for $1/\theta$ when $c_R/c_D = 0.1$ and $\lambda = 1$. This indicates that optimum N_i^* ($i = 3, 4, 5$) decreases with $1/\theta$ and $N_3^* \ge N_4^*$ and $N_5^* \ge N_3^*$. Compared to Table 5.4, $N_2^* \ge N_5^*$, however, two values are almost the same. In addition, when $1/\theta = 0.50$, $N_i^* = 1$ for all $i = 1, 2, 3, 4, 5$, and decrease to 1 with $1/\theta$. □

5.5 Checkpoint Models with Random Works

Suppose that we have to make the process of a job with N works each of which has a random time Y_j ($j = 1, 2, \ldots, N$) and executes successively until N works are completed. A double modular system for failure detection of each work is adopted [4, p. 124]: We place checkpoints at every completion of each work, which have two functions that store and compare the state of processes. To detect some errors in the process, we execute two independent modules and compare two states of modules at successive times S_j ($j = 1, 2, \ldots, N$). If two states match equally, then two modules are correct, and we proceed to the next interval. Conversely, if two states do not match, then it is judged that some errors in two modules have occurred. In this case, we go back to the newest checkpoint and restart again two modules. We repeat the above procedures until two states match for each checkpoint interval. The process of one work is completed successfully when two modules have been correct for all N intervals (Fig. 5.4).

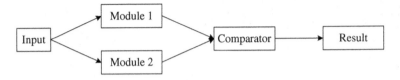

Fig. 5.4 Error detection by a double modular system

5.5.1 Three Schemes with Two Types of Checkpoints

Introduce two types of checkpoints: Compare-and-store-checkpoint (CSCP) which stores and compares the states of both processes, and compare-checkpoint (CCP), which compares only the states without storing them [5, 13, 14]. Using two checkpoints, we consider the following three checkpoint schemes:

(1) Scheme 1: CSCP is placed at time S_j $(j = 1, 2, \ldots, N)$.
(2) Scheme 2: CSCP is placed only at S_N.
(3) Scheme 3: CCP is placed at every end of S_j $(j = 1, 2, \ldots, N - 1)$ and CSCP is placed at S_N.

It is assumed that c is the overhead for comparison of two states and c_S is the overhead for their store, i.e., $c + c_S$ for CSCP and c for CCP. Then, the total working time until the completion of N works is obtained, using renewal equations used in Sect. 5.3. Three schemes are compared and the best scheme among them is determined numerically [5].

We consider the process of a job with N works in Sect. 5.3. In this case, to detect failures or errors, we provide two independent modules, where they can compare and store two states at checkpoint times. Some failures of one module occur at constant rate λ $(0 < \lambda < \infty)$, i.e., the probability that two modules have no failure in $[0, t]$ is $e^{-2\lambda t}$.

5.5.1.1 Scheme 1

CSCP is placed at every end of work j $(j = 1, 2, \ldots, N)$ in Fig. 5.5. When two states of modules match with each other at the end of work j, the process of work j is correct and its state is stored. In this case, two modules go ahead and execute work $(j + 1)$. However, when two states do not match, it is judged that some failures have occurred, and two modules go back and reexecute work j again. The process ends when two states of each module match at the end of work N, and its state is stored.

Noting that the system with two modules has an exponential failure distribution $(1 - e^{-2\lambda t})$ and Y_j has a general distribution with finite mean $1/\theta$ $(0 < \theta < \infty)$, the working time of work j is given by a renewal function

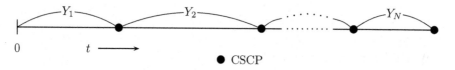

Fig. 5.5 Process of working times for Scheme 1

$$L_1(1) = \int_0^\infty \left\{ e^{-2\lambda t}(c + c_S + t) + (1 - e^{-2\lambda t})[c + t + L_1(1)] \right\} dG(t).$$

Solving this equation for $L_1(1)$,

$$L_1(1) = \frac{c + 1/\theta}{G^*(2\lambda)} + c_S, \qquad (5.45)$$

where $G^*(s) \equiv \int_0^\infty e^{-st} \, dG(t)$ for $Re(s) \geq 0$. Therefore, the total working time of N works is

$$L_1(N) = N \left[\frac{c + 1/\theta}{G^*(2\lambda)} + c_S \right] \quad (N = 1, 2, \ldots). \qquad (5.46)$$

5.5.1.2 Scheme 2

CSCP is placed only at the end of work N in Fig. 5.6: When two states of all work j $(j = 1, 2, \ldots, N)$ match at the end of work N, their states are stored. When two states of at least one work in N works do not match, two modules go back to work 1 and make their reexecutions again.

In a similar way of obtaining (5.45), the total working time of N works is

$$L_2(N) = \int_0^\infty \left\{ e^{-2\lambda t}(c + c_S + t) + (1 - e^{-2\lambda t})[c + t + L_2(N)] \right\} dG^{(N)}(t),$$

i.e.,

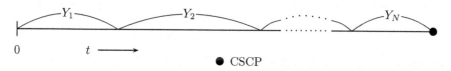

Fig. 5.6 Process of working times for Scheme 2

$$L_2(N) = \frac{c + N/\theta}{[G^*(2\lambda)]^N} + c_S \quad (N = 1, 2, \ldots). \tag{5.47}$$

We find an optimum number N_2^* which minimizes

$$\widehat{L}_2(N) = \frac{L_2(N)}{N} = \frac{1}{N} \left\{ \frac{c + N/\theta}{[G^*(2\lambda)]^N} + c_S \right\} \quad (N = 1, 2, \ldots). \tag{5.48}$$

From the inequality $\widehat{L}_2(N + 1) - \widehat{L}_2(N) \geq 0$,

$$\frac{c[N - (N + 1)G^*(2\lambda)] + N(N + 1)[1 - G^*(2\lambda)]/\theta}{[G^*(2\lambda)]^{N+1}} \geq c_S, \tag{5.49}$$

whose left-hand side increases strictly with N to ∞. Therefore, there exists a finite and unique minimum N_2^* $(1 \leq N_2^* < \infty)$ which satisfies (5.49).

5.5.1.3 Scheme 3

CSCP is placed only at the end of work N and CCP is placed at every end of work j $(j = 1, 2, \ldots, N - 1)$ between CSCPs in Fig. 5.7: When two states of work j $(j = 1, 2, \ldots, N - 1)$ match at the end of work j, two modules execute work $j + 1$. When two states of work j $(j = 1, 2, \ldots, N)$ do not match, two modules go back to work 1. When two states of work N match, the process is completed and its state is stored.

Let $\widetilde{L}_3(j)$ be working time from work j to the completion of work N. Then, by the similar method for obtaining (5.47),

$$\widetilde{L}_3(j) = \int_0^\infty \left\{ e^{-2\lambda t}[c + t + \widetilde{L}_3(j + 1)] + (1 - e^{-2\lambda t})[c + t + \widetilde{L}_3(1)] \right\} dG(t)$$

$$(j = 1, 2, \ldots, N - 1),$$

$$\widetilde{L}_3(N) = \int_0^\infty \left\{ e^{-2\lambda t}(c + c_S + t) + (1 - e^{-2\lambda t})[c + t + \widetilde{L}_3(1)] \right\} dG(t). \tag{5.50}$$

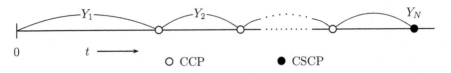

0          $t \longrightarrow$ ○ CCP   ● CSCP

Fig. 5.7 Process of working times for Scheme 2

Solving the above equation for $\widetilde{L}_3(1)$, the total working time of N works is (Problem 13 in Sect. 5.6)

$$L_3(N) \equiv \widetilde{L}_3(1) = \frac{(c + 1/\theta)\{1 - [G^*(2\lambda)]^N\}}{[1 - G^*(2\lambda)][G^*(2\lambda)]^N} + c_S \quad (N = 1, 2, \ldots). \quad (5.51)$$

When $N = 1$, all working times $L_i(N)$ $(i = 1, 2, 3)$ agree with each other.

We find an optimum number N_3^* which minimizes

$$\widehat{L}_3(N) \equiv \frac{L_3(N)}{N} = \frac{1}{N} \left(\frac{(c + 1/\theta)\{1 - [G^*(2\lambda)]^N\}}{[1 - G^*(2\lambda)][G^*(2\lambda)]^N} + c_S \right) \quad (N = 1, 2, \ldots).$$

$$(5.52)$$

From the inequality $\widehat{L}_3(N + 1) - \widehat{L}_3(N) \geq 0$,

$$\frac{N - (N + 1)G^*(2\lambda)}{[G^*(2\lambda)]^{N+1}} + 1 \geq \frac{c_S[1 - G^*(2\lambda)]}{c + 1/\theta}, \quad (5.53)$$

whose left-hand side increases strictly with N to ∞. Thus, there exists a finite and unique minimum N_3^* $(1 \leq N_3^* < \infty)$ which satisfies (5.53). Therefore, it would be better to make the process with N_2^* and N_3^* for Schemes 2 and 3, respectively.

5.5.2 Comparison of Three Schemes

We compare three working times $L_i(N)$ in (5.46), (5.47), and (5.51) for $N \geq 2$:

(i) $\quad \dfrac{c/N + 1/\theta}{[G^*(2\lambda)]^N} + \dfrac{c_S}{N} \geq \dfrac{c + 1/\theta}{G^*(2\lambda)} + c_S \Rightarrow L_2(N) \geq L_1(N).$

(ii) $\quad \dfrac{\sum_{j=0}^{N-2}\{[G^*(2\lambda)]^j - [G^*(2\lambda)]^{N-1}\}}{(N - 1)[G^*(2\lambda)]^N} \geq \dfrac{c_S}{c + 1/\theta} \Rightarrow L_3(N) \geq L_1(N).$

(iii) $\quad \dfrac{1}{N - 1} \sum_{j=1}^{N-1}[G^*(2\lambda)]^j \geq \dfrac{1/\theta}{c + 1/\theta} \Rightarrow L_3(N) \geq L_2(N).$

Note that case (iii) does not depend on c_S and $[1/(N-1)] \sum_{j=1}^{N-1}[G^*(2\lambda)]^j$ decreases strictly from $G^*(2\lambda)$ to 0 for $N \geq 2$. Thus, there exists a finite and unique minimum N_3 $(2 \leq N_3 < \infty)$ which satisfies

$$\frac{1}{N - 1} \sum_{j=1}^{N-1}[G^*(2\lambda)]^j \leq \frac{1}{c\theta + 1}. \quad (5.54)$$

Table 5.7 Bound numbers N_i ($i = 1, 2, 3$) when $G(t) = 1 - e^{-\theta t}$ and $c_s\theta = 0.1$

λ/θ	$c\theta = 0.1$			$c\theta = 0.5$		
	N_1	N_2	N_3	N_1	N_2	N_3
0.001	92	86	97	235	64	438
0.005	19	18	20	47	13	88
0.010	10	9	10	24	7	44
0.050	2	2	2	5	2	10
0.100	2	2	2	3	2	5
0.500	2	2	2	2	2	2

Thus, if $N < N_3$, then $L_3(N) > L_2(N)$, and conversely, if $N \geq N_3$, then $L_3(N) \leq L_2(N)$. That is, if a job consists of works with some large size, Scheme 3 is better than Scheme 2. When $G(t) = 1 - e^{-\theta t}$, (5.54) is

$$\frac{1}{N-1} \sum_{j=1}^{N-1} \left(\frac{1}{2\lambda/\theta + 1}\right)^j \leq \frac{1}{c\theta + 1}. \tag{5.55}$$

Clearly, N_3 increases with $c\theta$ and decreases with λ/θ. Similarly, we compute bound numbers N_1 for case (i) and N_2 for case (ii).

Example 5.7 (*Checking number for exponential failure time*) Table 5.7 presents the bound numbers N_i ($i = 1, 2, 3$) for λ/θ when $c\theta = 0.1, 0.5$. For example, when $\lambda/\theta = 0.1$ and $c\theta = 0.5$, Scheme 3 is better than Scheme 2 for $N \geq 5$. In this case, when $N = 4$, Scheme 1 is better than Scheme 2 and Scheme 2 is better than Scheme 3. When $c\theta = 0.1$, all N_i ($i = 1, 2, 3$) are almost the same, because three schemes are changed little in the case where c is small. When $\lambda/\theta = 0.5$, $N_i = 2$ ($i = 1, 2, 3$), and hence, all schemes are the same.   □

5.5.3 Comparison of Four Works

We consider the following six schemes of a job with four works, where checkpoints are placed at the completion of work 4, and (i) No checkpoint at $j = 1, 2, 3$, (ii) CCP at $j = 2$, (iii) CSCP at $j = 2$, (iv) CCP at $j = 1, 3$, CSCP at $j = 2$, (v) CCP at $j = 1, 2, 3$, and (vi) CSCP at $j = 1, 2, 3$ in Fig. 5.8.

By similar methods for obtaining the previous three schemes, the total mean execution times L_i ($i = 1, 2, 3, 4, 5, 6$) are obtained as follows:

Fig. 5.8 Process of check-points for four works

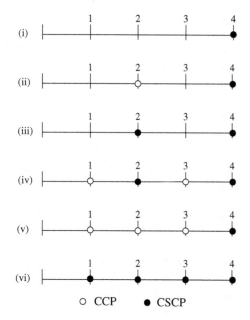

○ CCP   ● CSCP

(i) $L_1 = \dfrac{c + 4/\theta}{[G^*(2\lambda)]^4} + c_S.$

(ii) $L_2 = \dfrac{(c + 2/\theta)\{1 + [G^*(2\lambda)]^2\}}{[G^*(2\lambda)]^4} + c_S.$

(iii) $L_3 = 2\left\{\dfrac{c + 2/\theta}{[G^*(2\lambda)]^2} + c_S\right\}.$

(iv) $L_4 = 2\left\{\dfrac{(c + 1/\theta)[1 + G^*(2\lambda)]}{[G^*(2\lambda)]^2} + c_S\right\}.$

(iv) $L_5 = \dfrac{(c + 1/\theta)[1 + G^*(2\lambda)]\{1 + [G^*(2\lambda)]^2\}}{[G^*(2\lambda)]^4} + c_S.$

(vi) $L_6 = 4\left[\dfrac{c + 1/\theta}{G^*(2\lambda)} + c_S\right].$

Example 5.8 (*Optimum scheme for four works*) Tables 5.8 and 5.9 present the total working times $L_i\theta$ ($i = 1, 2, 3, 4, 5, 6$) when $c\theta = 0.1, 0.5$, $c_S\theta = 0.1$, where the asterisk shows the best scheme among 6 ones for λ/θ. This indicates that Scheme 1 becomes better as λ/θ is smaller and Scheme 6 becomes better as λ/θ is larger. If $\lambda/\theta < 0.058$ in Table 5.8 and $\lambda/\theta < 0.023$ in Table 5.9, then Scheme 1 is better than Scheme 6. That is, when λ/θ is large, we should place many CSCP. In this example, we do not place any CCP. However, if c is smaller compared with c_S and the number of works is larger, Schemes 2, 4, and 5 might be better than other schemes (Problem 14 in Sect. 5.6). □

Table 5.8 Mean execution times $L_i\theta$ ($i = 1, 2, 3, 4, 5, 6$) when $G(t) = 1 - e^{-\theta t}$, $c\theta = 0.5$, and $c_S\theta = 0.1$

λ/θ	$L_1\theta$	$L_2\theta$	$L_3\theta$	$L_4\theta$	$L_5\theta$	$L_6\theta$
0.001	4.636[a]	5.130	5.220	6.218	6.130	6.412
0.005	4.783[a]	5.252	5.301	6.290	6.252	6.460
0.010	4.971[a]	5.407	5.402	6.381	6.406	6.520
0.050	6.688	6.785	6.250[a]	7.130	7.758	7.000
0.100	9.431	8.884	7.400[a]	8.120	9.762	7.600
0.500	72.100	50.100	20.200	18.200	45.100	12.400[a]
1.000	364.600	225.100	45.200	36.200	180.100	18.400[a]

[a] Minimum value

Table 5.9 Mean execution times $L_i\theta$ ($i = 1, 2, 3, 4, 5, 6$) when $G(t) = 1 - e^{-\theta t}$, $c\theta = 0.1$, and $c_S\theta = 0.1$

λ/θ	$L_1\theta$	$L_2\theta$	$L_3\theta$	$L_4\theta$	$L_5\theta$	$L_6\theta$
0.001	4.233[a]	4.325	4.417	4.613	4.522	4.809
0.005	4.366[a]	4.427	4.484	4.666	4.611	4.844
0.010	4.538[a]	5.558	4.570	4.733	4.724	4.888
0.050	6.103	5.716	5.282	5.282	5.716	5.240[a]
0.100	8.602	7.479	6.248	6.008	7.186	5.680[a]
0.500	65.700	42.100	17.000	13.400	33.100	9.200[a]
1.000	332.200	189.100	38.000	26.600	132.100	13.600[a]

[a] Minimum value

5.6 Problems

1 Prove that (5.1) + (5.2) = 1.
2 Derive (5.4) and (5.5) from (5.3).
3 Make certain and discuss generally that when the random checking cost is the half of the periodic one, both expected costs are almost the same.
4 Prove that T_P^* decreases with $1/\theta$ to T^*.
5 Derive (5.20) from (5.19) and (5.21) from (5.20).
6 Prove that N_2^* decreases with $1/\theta$ from ∞ to 1.
7 Consider the reason why two optimum policies which minimize $C_{R1}(N)$ in (5.20) and $C_{R3}(N)$ in (5.30) are coincident with each other.
8 Derive $C_R(i)$ ($i = 1, 2, 3, 4$).
9 Derive (5.31).
10 Prove that the left-hand side of (5.34) increases strictly with N to ∞.
11 Prove that $N_4^* \le N_3^*$.
12 Prove that $N_5^* \ge N_3^*$.
13 Derive (5.51) from (5.50).
14 Consider 6 works with two checkpoints and compare them numerically.

References

1. Reuter A (1984) Performance analysis of recovery techniques. ACM Trans Database Syst 9:526–559
2. Fukumoto S, Kaio N, Osaki S (1992) A study of checkpoint generations for a database recovery mechanism. Comput Math Appl 24:63–70
3. Naruse K, Nakagawa S, Okuda Y (2007) Optimal checking time of backup operation for a database system. In: Dohi T, Osaki S, Sawaki K (eds) Recent advances in stochastic operations research. World Scientific, Singapore, pp 131–144
4. Nakagawa T (2009) Advanced reliability models and maintenance policies. Springer, London
5. Nakagawa T, Naruse K, Maeji S (2009) Random checkpoint models with N tandem tasks. IEICE Trans Fundam E92–A:1572–1577
6. Naruse K, Nakagawa T, Maeji S (2009) Random checking times of backup operation for a database system. 15th ISSAT International Conference on Reliab and Qual in Design, pp 339–342.
7. Naruse K, Nakagawa T, Maeji S (2010) Optimal checking models with random working times. Proceedings of 4th Asia-Pacific International Symposium (APARM 2010), pp 488–495.
8. Siewiorek DP, Swarz R (1983) The theory and practice of reliable system design. Digital Press, Bedford
9. Lee PA, Anderson T (1990) Fault tolerance-principles and practice. Springer, Wien
10. Nakagawa S, Fukumoto S, Ishii N (2003) Optimal checkpointing intervals of three error detection scheme by a double modular redundancy. Math Comput Model 38:1357–1363
11. Pradhan DK, Vaidya NH (1994) Roll-forward and rollback recovery: performance-reliability trade-off. In: 24th international symposium on fault-tolerant computing, pp 186–195.
12. Nakagawa S, Okuda Y, Yamada S (2003) Optimal checkpointing interval for task duplication with spare processing. 9th ISSAT International conference on Reliab and Qual in Design, pp 215–219.
13. Ziv A, Bruck J (1997) Performance optimization of checkpointing schemes with task duplication. IEEE Trans Comput 46:1381–1386
14. Ziv A, Bruck J (1998) Analysis of checkpointing schemes with task duplication. IEEE Trans Comput 47:222–227
15. Kim H, Shin KG (1996) Design and analysis of an optimal instruction-retry policy for TMR controller computers. IEEE Trans Comput 45:1217–1225
16. Ohara M, Suzuki R, Arai M, Fukumoto S, Iwasaki K (2006) Analytical model on hybrid state saving with a limited number of checkpoints and bound rollbacks. IEICE Trans Fundam E89–A:2386–2395
17. Nakagawa T (2011) Stochastic processes with applications to reliability theory. Springer, London
18. Munford AG (1981) Comparison among certain inspection policies. Manage Sci 27:260–267

Chapter 6
Random Parallel Systems

High system reliability can be achieved by redundancy and maintenance. The most typical model is a standard parallel system, which consists of n identical units in parallel. It was originally shown that the system can operate for a specified mean time by either changing the replacement time or increasing the number of units [1, p. 65]. The reliabilities of many redundant systems were computed and summarized [2]. A variety of redundant systems with multiple failure modes and their optimization problems were discussed [3]. Reliabilities of parallel and parallel-series systems with dependent failures of components were derived [4]. Some optimization methods of redundancy allocation for series-parallel systems were studied [5]. A good survey of multistate and consecutive k-out-of-n systems was done [6, 7].

A variety of optimization problems encountered in redundancy were summarized [8, p. 7]: Optimum number of units and replacements times were derived, and several applications to redundant data transmissions, bits, and networks were shown. Recently, a simple asymptotic method of computing the mean time to system failure was given [9], and optimization problems of a parallel system with random number of units were firstly studied [10].

This chapter attempts to apply the results in Chap. 2 [11–13] to a parallel system with n units: First, we summarize the computation results of the mean time to system failure and their approximations when the failure time has a Weibull distribution. Furthermore, when the number of units is a random variable with a Poisson distribution, the MTTF is computed.

Next, the basic problems in a parallel system are that how many number of units is optimum and when the system should be replaced before failure from economical viewpoints. When the number of units is constant and random, an optimum number of units and replacement time which minimize the expected cost rates are derived analytically and numerically. Similar discussions of a K-out-of-n system when K is constant and random are made.

Finally, using shortage and excess costs introduced in Sect. 2.1.1, optimum replacement time and random replacement time which minimize the expected costs are discussed. When the number of units is random, an optimum replacement time is also obtained. Furthermore, when the system operates for a job with random working

© Springer-Verlag London 2014
T. Nakagawa, *Random Maintenance Policies*,
Springer Series in Reliability Engineering, DOI 10.1007/978-1-4471-6575-0_6

times, an optimum replacement number of working times and an optimum number
of units will be derived in Chap. 7.

Suppose that the failure time of each unit has an identical distribution $F(t)$ with
finite mean $\mu \equiv \int_0^\infty \overline{F}(t)\, dt < \infty$, where $\overline{\Phi}(t) \equiv 1 - \Phi(t)$. When $F(t)$ has a
density function $f(t) \equiv dF(t)/dt$, the failure rate $h(t) \equiv f(t)/\overline{F}(t)$ increases to
$h(\infty) \equiv \lim_{t\to\infty} h(t)$.

6.1 MTTF of a Parallel System

Consider a standard parallel system with n $(n = 1, 2, \ldots)$ identical units where each
unit is independent and has a failure distribution $F(t)$ with mean μ $(0 < \mu < \infty)$.
Then, the mean time to system failure (MTTF) is

$$\mu_n = \int_0^\infty \left[1 - F(t)^n\right] dt \quad (n = 1, 2, \ldots), \tag{6.1}$$

where $\mu_1 \equiv \mu$. In particular, when $F(t) = 1 - e^{-\lambda t}$ $(0 < \lambda < \infty)$, the MTTF is
(Problem 1 in Sect. 6.5)

$$\mu_n = \int_0^\infty \left[1 - (1 - e^{-\lambda t})^n\right] dt = \frac{1}{\lambda}\sum_{j=1}^n \frac{1}{j}. \tag{6.2}$$

Note that $H_n \equiv \sum_{j=1}^n (1/j)$ is called a harmonic series of a natural number, and
increases strictly with n very slowly and logarithmically to ∞ [14]: In addition, H_n
has the following approximations:

$$\log n + \frac{1}{n} < H_n < \log n + 1 \quad (n = 2, 3, \ldots),$$
$$H_n \approx \gamma + \log n, \tag{6.3}$$

where γ is Euler's constant and $\gamma \equiv 0.5772156619\ldots$.

Example 6.1 (MTTF for exponential failure time) Table 6.1 presents the exact MTTF
in (6.2), approximations of $\log n + 1/n$, $\log n + 1$, and $\log n + \gamma$ when $F(t) = 1 - e^{-t}$.
This indicates that the three approximations give good ones to exact MTTF, and
$\log n + 1/n < \log n + \gamma <$ MTTF $< \log n + 1$ for $n \geq 2$. Specially, $\log n + \gamma$ is
much better for large n and gives a lower bound for the exact MTTF. $\square$

Table 6.1 Exact MTTF and three approximations when $F(t) = 1 - e^{-t}$

n	MTTF	$\log n + \frac{1}{n}$	$\log n + 1$	$\log n + \gamma$
1	1.00000	1.00000	1.00000	0.57722
2	1.50000	1.19315	1.69315	1.27036
3	1.83333	1.43195	2.09861	1.67583
4	2.08333	1.63629	2.38629	1.96351
5	2.28333	1.80944	2.60944	2.18665
6	2.45000	1.95843	2.79176	2.36898
7	2.59286	2.08877	2.94591	2.52313
8	2.71786	2.20444	3.07944	2.65666
9	2.82897	2.30834	3.19722	2.77444
10	2.92897	2.40259	3.30259	2.87980
20	3.59774	3.04573	3.99573	3.57295
30	3.99499	3.43453	4.40120	3.97841
40	4.27854	3.71388	4.68888	4.26610
50	4.49921	3.93202	4.91202	4.48924
60	4.67987	4.11101	5.09434	4.67156
70	4.83284	4.26278	5.24850	4.82571
80	4.96548	4.39453	5.38203	4.95924
90	5.08257	4.51092	5.49981	5.07703
100	5.18738	4.61517	5.60517	5.18239

6.1.1 Asymptotic Methods

It is assumed that the failure time X of each unit has a general distribution $F(t) \equiv \Pr\{X \le t\} = 1 - e^{-H(t)}$ with finite mean $\mu \equiv \int_0^\infty e^{-H(t)} dt$, where $H(t) \equiv \int_0^t h(u) du$ is called the *cumulative hazard rate* denoted in (1.10). A random variable $Y \equiv H(X)$ has the following distribution [16, p. 6]:

$$\Pr\{Y \le t\} = \Pr\{H(X) \le t\} = \Pr\{X \le H^{-1}(t)\} = 1 - e^{-t}, \quad (6.4)$$

where H^{-1} is the inverse function of H. Thus, Y has an exponential distribution with mean 1, and $E\{H(X)\} = 1$. Making a Taylor's series expansion to the first order of $\mu = E\{X\}$,

$$Y = H(X) \approx H(\mu) + (X - \mu)H'(\mu).$$

Thus, the expectation of Y is approximately

$$E\{Y\} = E\{H(X)\} \approx H(E\{X\}),$$

i.e.,

$$E\{X\} \approx H^{-1}(E\{Y\}). \tag{6.5}$$

Let Y_n ($n = 1, 2, \ldots$) be the exponential failure times with mean 1 of an n-unit parallel system and X_n ($n = 1, 2, \ldots$) be the failure times of the same system in which each unit has a failure distribution $F(t) = 1 - e^{-H(t)}$. Then, because $E\{Y_n\} = \sum_{j=1}^{n}(1/j)$ from (6.2), and hence, from (6.5),

$$E\{X_n\} \approx H^{-1}\left(\sum_{j=1}^{n}\frac{1}{j}\right).$$

In particular, when each unit has the same Weibull failure distribution, i.e., $H(t) = t^\alpha$ ($\alpha > 0$) and its mean is $\Gamma(1 + 1/\alpha) = \Gamma(1/\alpha)/\alpha$, where $\Gamma(\alpha + 1) \equiv \int_0^\infty x^\alpha e^{-x}\, dx$ ($\alpha > 0$),

$$E\{X_n\} \approx \left(\sum_{j=1}^{n}\frac{1}{j}\right)^{1/\alpha} \quad (n = 1, 2, \ldots). \tag{6.6}$$

It is noted that from Jensen's inequality,

$$\text{MTTF} = \int_0^\infty \left[1 - (1 - e^{-t^\alpha})\right] dt \quad \begin{cases} < \left(\sum_{j=1}^{n}\frac{1}{j}\right)^{1/\alpha} & \text{for } \alpha > 1, \\ = \left(\sum_{j=1}^{n}\frac{1}{j}\right) & \text{for } \alpha = 1, \\ > \left(\sum_{j=1}^{n}\frac{1}{j}\right)^{1/\alpha} & \text{for } \alpha < 1. \end{cases} \tag{6.7}$$

Example 6.2 (MTTF for Weibull failure time) Tables 6.2 and 6.3 present exact MTTF, three approximations of $[\sum_{j=1}^{n}(1/j)]^{1/\alpha}$, $[\Gamma(1 + 1/\alpha) + \sum_{j=2}^{n}(1/j)]^{1/\alpha}$, and $(\log n + \gamma)^{1/\alpha}$ for $\alpha = 2, 3$. These approximations give good ones to exact MTTF for any $n \geq 2$, and MTTF $< (\log n + \gamma)^{1/\alpha} < \sum_{j=1}^{n}(1/j)^{1/\alpha}$ for $n \geq 3$. It would be sufficient in practical fields to estimate $[\Gamma(1 + 1/\alpha) + \sum_{j=2}^{n}(1/j)]^{1/\alpha}$ as a lower bound and $(\log n + \gamma)^{1/\alpha}$ as an upper bound for large n. □

We try to apply the above results to the unit which undergoes minimal repair at failures in Chap. 3, i.e., failures occur at a nonhomogeneous Poisson process with with mean value function $H(t)$ [15, p. 62]. Letting $N(t)$ be the number of failures in $[0, t]$ [16, p. 97],

$$\Pr\{N(t) = k\} = \frac{[H(t)]^k}{k!} e^{-H(t)} \quad (k = 0, 1, 2, \ldots),$$

$$E\{N(t)\} = H(t). \tag{6.8}$$

Furthermore, let S_k be the successive failure times of the unit and $X_k \equiv S_k - S_{k-1}$ ($k = 1, 2, \ldots$) be the times between failures with $F(t) = \Pr\{X_1 \leq t\} = $

Table 6.2 Exact MTTF and three approximations when $\alpha = 2$

n	MTTF	$(\log n + \gamma)^{1/2}$	$[\sum_{j=1}^{n}(1/j)]^{1/2}$	$[\Gamma(1+1/2) + \sum_{j=2}^{n}(1/j)]^{1/2}$
1	0.88623	0.75975	1.00000	0.88623
2	1.14580	1.12710	1.22474	1.17738
3	1.29037	1.29454	1.35401	1.31132
4	1.38851	1.40125	1.44338	1.40341
5	1.46196	1.47873	1.51107	1.47294
6	1.52027	1.53915	1.56525	1.52847
7	1.56838	1.58844	1.61024	1.57451
8	1.60921	1.62993	1.64859	1.61372
9	1.64458	1.66567	1.68195	1.64778
10	1.67572	1.69700	1.71142	1.67785
20	1.86977	1.89022	1.89677	1.86654
30	1.97542	1.99460	1.99875	1.97008
40	2.04729	2.06545	2.06846	2.04078
50	2.10142	2.11878	2.12113	2.09414
60	2.14467	2.16138	2.16330	2.13684
70	2.18058	2.19675	2.19837	2.17234
80	2.21122	2.22694	2.22834	2.20266
90	2.23791	2.25323	2.25446	2.22908
100	2.26151	2.27649	2.27758	2.25247

Table 6.3 Exact MTTF and three approximations when $F(t) = 1 - e^{-t^3}$

n	MTTF	$(\log n + \gamma)^{1/3}$	$[\sum_{j=1}^{n}(1/j)]^{1/3}$	$[\Gamma(1+1/3) + \sum_{j=2}^{n}(1/j)]^{1/3}$
1	0.89298	0.83262	1.00000	0.89298
2	1.07720	1.08304	1.14471	1.11682
3	1.17182	1.18780	1.22390	1.19961
4	1.23346	1.25221	1.27718	1.25493
5	1.27840	1.29796	1.31681	1.29591
6	1.31340	1.33307	1.34810	1.32818
7	1.34188	1.36138	1.37381	1.35464
8	1.36577	1.38499	1.39554	1.37698
9	1.38626	1.40516	1.41430	1.39624
10	1.40417	1.42272	1.43078	1.41313
20	1.51292	1.52877	1.53230	1.51695
30	1.57031	1.58454	1.58674	1.57244
40	1.60869	1.62185	1.62342	1.60977
50	1.63727	1.64965	1.65087	1.63767
60	1.65991	1.67168	1.67267	1.65982
70	1.67858	1.68987	1.69070	1.67813
80	1.69443	1.70532	1.70603	1.69369
90	1.70816	1.71871	1.71934	1.70718
100	1.72026	1.73052	1.73108	1.71909

$1 - e^{-H(t)}$, where $S_0 \equiv 0$. Then, we have [16, p. 97]

$$E\{X_k\} = \int_0^\infty \frac{[H(t)]^{k-1}}{(k-1)!} e^{-H(t)} dt \quad (k = 1, 2, \ldots),$$

$$E\{S_n\} = \sum_{k=0}^{n-1} \int_0^\infty \frac{[H(t)]^k}{k!} e^{-H(t)} dt \quad (n = 1, 2, \ldots). \tag{6.9}$$

Letting $H(t_k) = k$, t_k represents the time that the expected number of failures is k from (6.8). Thus, when $x_k = t_k - t_{k-1}$, $H(x_k + t_{k-1}) - H(t_{k-1}) = 1$, which represents that the expected number of failures in $[t_{k-1}, t_{k-1} + x_k]$ is 1. It is assumed that the failure time has a Weibull distribution, i.e., $F(t) = 1 - \exp(-t^\alpha)$ and $H(t) = t^\alpha$ $(\alpha > 0)$. Then, from (6.9) (Problem 2 in Sect. 6.5), the mean time between failures (MTBF) is

$$E\{X_k\} = \frac{1}{\alpha} \frac{\Gamma(k - 1 + 1/\alpha)}{(k - 1)!} \quad (k = 1, 2, \ldots), \tag{6.10}$$

and the mean time to the nth failure is

$$E\{S_n\} = \frac{\Gamma(n + 1/\alpha)}{(n - 1)!} \quad (n = 1, 2, \ldots). \tag{6.11}$$

Furthermore, when $t_k^\alpha = k$, i.e., $t_k = k^{1/\alpha}$,

$$x_k = t_k - t_{k-1} = k^{1/\alpha} - (k - 1)^{1/\alpha} \quad (k = 1, 2, \ldots). \tag{6.12}$$

Example 6.3 (MTBF for Weibull failure time) Table 6.4 presents exact $E\{X_k\}$ in (6.10) and x_k in (6.12) for $\alpha = 1.5, 2.0, 2.5, 3.0$. This indicates that x_k becomes better approximations to the exact MTBF as k becomes larger for any α, and such approximations become better as α becomes smaller. When $\alpha = 1$, $E\{X_k\} = x_k = 1$. It is of interest that $E\{X_1\} < x_1$, however, $E\{X_k\} > x_k$ for $k \geq 2$. This indicates that it would be easy to estimate MTBF when failures occur at a nonhomogeneous Poisson process with a Weibull distribution.   □

6.1.2 Random Number of Units

It has been assumed until now that the number n of units for a parallel system is constant and is previously known. However, when n is large, it might be sometimes encountered with the case where we could not know the exact number of units and

Table 6.4 Exact MTBF and approximations when $F(t) = 1 - e^{-t^\alpha}$

k	$\alpha = 1.5$		$\alpha = 2.0$		$\alpha = 2.5$		$\alpha = 3.0$	
	$E\{X_k\}$	x_k	$E\{X_k\}$	x_k	$E\{X_k\}$	x_k	$E\{X_k\}$	x_k
1	0.9027	1.0000	0.8862	1.0000	0.8873	1.0000	0.8930	1.0000
2	0.6018	0.5874	0.4431	0.4142	0.3549	0.3195	0.2977	0.2599
3	0.5015	0.4927	0.3323	0.3178	0.2484	0.2323	0.1984	0.1823
4	0.4458	0.4392	0.2769	0.2679	0.1987	0.1893	0.1543	0.1452
5	0.4087	0.4042	0.2423	0.2361	0.1689	0.1626	0.1286	0.1226
6	0.3814	0.3779	0.2181	0.2134	0.1487	0.1440	0.1115	0.1071
7	0.3602	0.3574	0.1999	0.1963	0.1338	0.1302	0.0991	0.0958
8	0.3431	0.3407	0.1856	0.1827	0.1223	0.1195	0.0896	0.0871
9	0.3288	0.3267	0.1740	0.1716	0.1132	0.1108	0.0822	0.0801
10	0.3166	0.3148	0.1644	0.1623	0.1056	0.1037	0.0761	0.0744
20	0.2484	0.2477	0.1140	0.1132	0.0679	0.0673	0.0465	0.0460
30	0.2162	0.2158	0.0924	0.0921	0.0528	0.0525	0.0352	0.0349
40	0.1960	0.1958	0.0798	0.0796	0.0443	0.0441	0.0289	0.0287
50	0.1818	0.1816	0.0712	0.0711	0.0386	0.0385	0.0248	0.0247
100	0.1439	0.1439	0.0502	0.0501	0.0254	0.0253	0.0156	0.0155

estimate statistically only its probability distribution because objective systems are complex, big or old.

Consider a parallel system with N units in which N is a random variable with a Poisson distribution with mean β ($\beta \geq 1$) such that

$$\Pr\{N = n\} = \frac{\beta^n}{n!} e^{-\beta} \quad (n = 0, 1, 2, \ldots). \tag{6.13}$$

Then, the reliability of the system at time t is

$$R_\beta(t) = \sum_{n=1}^{\infty} \left[1 - F(t)^n\right] \frac{\beta^n}{n!} e^{-\beta} = 1 - e^{-\beta \overline{F}(t)}. \tag{6.14}$$

Note that $R_\beta(t)$ arises when a parallel system with zero unit is considered as a degenerated system with a distribution function degenerated at time $t = 0$, i.e., the system has always failed for $t \geq 0$. Of course, the probability $\Pr\{N = 0\} = e^{-\beta}$ would be very small and could be neglected in actual fields.

In particular, when $\overline{F}(t) = e^{-\lambda t}$ ($0 < \lambda < \infty$),

$$R_\beta(t) = 1 - \exp(-\beta e^{-\lambda t}),$$

which is wellknown as Type I extreme distribution [16, p. 16]. Thus, because $\int_0^\infty [1 - \exp(e^{-\lambda t})] \, dt = \gamma/\lambda$ [17, p. 12], the MTTF is

$$\int_0^\infty [1 - \exp(-\beta e^{-\lambda t})] \, dt = \frac{1}{\lambda} (\log \beta + \gamma), \tag{6.15}$$

which agrees with (6.3) when $\beta = n$ and $\lambda = 1$.

6.2 Number of Units and Replacement Time

We discuss optimum number of units and replacement time for a parallel system with constant and random number of units.

6.2.1 Optimum Number of Units

We derive an optimum number n^* of units for a parallel system. From economical viewpoints, the expected cost rate is [8, p. 8], from (6.1),

$$C_1(n) = \frac{c_1 n + c_F}{\mu_n} \quad (n = 1, 2, \ldots), \tag{6.16}$$

where $c_1 = $ acquisition cost for one unit and $c_F = $ replacement cost for a failed system. We find an optimum number n^* which minimizes $C_1(n)$. Forming the inequality $C_1(n+1) - C_1(n) \geq 0$,

$$\frac{\mu_n}{\mu_{n+1} - \mu_n} - n \geq \frac{c_F}{c_1}, \tag{6.17}$$

whose left-hand side increases strictly to ∞. Therefore, there exists a finite and unique minimum n^* ($1 \leq n^* < \infty$) which satisfies (6.17).

In particular, when $F(t) = 1 - e^{-\lambda t}$, from (6.2) and (6.16), the expected cost rate is

$$\frac{C_1(n)}{\lambda} = \frac{c_1 n + c_F}{\sum_{j=1}^n (1/j)}, \tag{6.18}$$

and (6.17) becomes

$$\sum_{j=1}^n \frac{n+1}{j+1} \geq \frac{c_F}{c_1}, \tag{6.19}$$

whose left-hand side increases strictly from 1 to ∞. If $c_1 \geq c_F$, then $n^* = 1$.

Table 6.5 Optimum n^* and asymptotic $\tilde{n}$ when $F(t) = 1 - e^{-t^\alpha}$

c_F/c_1	$\alpha = 1$		$\alpha = 2$		$\alpha = 3$	
	n^*	$\tilde{n}$	n^*	$\tilde{n}$	n^*	$\tilde{n}$
1	1	2	1	1	1	1
2	2	3	1	2	1	1
5	4	5	2	3	2	2
10	6	7	3	4	3	3
20	10	10	6	6	4	4
50	19	19	10	10	8	7
100	32	32	17	17	13	12

Furthermore, using the approximation (6.3) for a harmonic series, the expected cost rate in (6.18) is approximately given by

$$\frac{\tilde{C}_1(n)}{\lambda} = \frac{c_1 n + c_F}{\log n + \gamma},$$ (6.20)

and an optimum $\tilde{n}$ to minimize $\tilde{C}_1(n)$ satisfies

$$\frac{\log n + \gamma}{\log(1 + 1/n)} - n \geq \frac{c_F}{c_1}.$$ (6.21)

In addition, using the approximation $\log(1 + x) \approx x$ for small x, (6.21) becomes approximately

$$n[\log n - (1 - \gamma)] \geq \frac{c_F}{c_1}.$$ (6.22)

Example 6.4 (Number of units for Weibull distribution failure time) Suppose that $F(t) = 1 - \exp(-t^\alpha)$ ($\alpha \geq 1$). Then, MTTF is given in (6.7), and from (6.17), an optimum n^* is given by a unique minimum such that

$$\frac{\int_0^\infty [1 - (1 - e^{-t^\alpha})^n] \, dt}{\int_0^\infty (1 - e^{-t^\alpha})^n e^{-t^\alpha} \, dt} - n \geq \frac{c_F}{c_1},$$ (6.23)

and from Example 6.2, an asymptotic number $\tilde{n}$ is

$$\frac{(\log n + \gamma)^{1/\alpha}}{[\log(n + 1) + \gamma]^{1/\alpha} - (\log n + \gamma)^{1/\alpha}} - n \geq \frac{c_F}{c_1}.$$ (6.24)

When $\alpha = 1$, (6.23) and (6.24) become (6.19) and (6.21), respectively. Table 6.5 presents optimum n^* and asymptotic $\tilde{n}$ for $\alpha = 1, 2, 3$ when $F(t) = 1 - \exp(-t^\alpha)$. Because $\sum_{j=1}^{n}(1/j)[(n + 1)/(j + 1)] \geq n$ when $\alpha = 1$, $n^* \leq c_F/c_1$. $\square$

Next, when the number N of units is a random variable with a Poisson distribution in (6.13), the expected cost rate in (6.16) is, from (6.14),

$$C_2(\beta) = \frac{c_1\beta + c_F}{\int_0^\infty [1 - e^{-\beta\overline{F}(t)}]\,dt}. \tag{6.25}$$

Differentiating $C_2(\beta)$ with respect to β and setting it equal to zero,

$$\frac{\int_0^\infty [1 - e^{-\beta\overline{F}(t)}]\,dt}{\int_0^\infty \overline{F}(t)e^{-\beta\overline{F}(t)}\,dt} - \beta = \frac{c_F}{c_1}, \tag{6.26}$$

whose left-hand side increases strictly with β from 0 to ∞ (Problem 3 in Sect. 6.5). Therefore, there exists a finite and unique β^* ($0 < \beta^* < \infty$) which satisfies (6.26). In particular, when $F(t) = 1 - e^{-\lambda t}$, (6.25) is

$$\frac{C_2(\beta)}{\lambda} = \frac{c_1\beta + c_F}{\log\beta + \gamma}. \tag{6.27}$$

Differentiating $C_2(\beta)$ with respect to β and setting it equal to zero,

$$\beta[\log\beta - (1 - \gamma)] = \frac{c_F}{c_1}, \tag{6.28}$$

whose left-hand side agrees with (6.22) when $\beta = n$, and increases strictly from 0 to ∞ for $\log\beta > 1 - \gamma$, i.e., $\beta > 1.53$. Therefore, there exists a finite and unique β^* ($1.53 < \beta^* < \infty$) which satisfies (6.28). In addition, setting that

$$x[\log x - (1 - \gamma)] = x,$$

$\log x = 2 - \gamma$, i.e., $x = e^{2-\gamma} \approx 4.15$. Thus, if $c_F/c_1 < 4.15$, then $\beta^* > c_F/c_1$, and conversely, if $c_F/c_1 \geq 4.15$, then $\beta^* \leq c_F/c_1$.

Example 6.5 (Random number of units for Weibull failure time) Suppose that $F(t) = 1 - \exp(-t^\alpha)$ ($\alpha \geq 1$). Then, the expected cost rate in (6.25) is

$$C_2(\beta) = \frac{c_1\beta + c_F}{\int_0^\infty [1 - \exp(-\beta e^{-t^\alpha})]\,dt}, \tag{6.29}$$

and (6.26) is

$$\frac{\int_0^\infty [1 - \exp(-\beta e^{-t^\alpha})]\,dt}{\int_0^\infty \exp(-t^\alpha - \beta e^{-t^\alpha})\,dt} - \beta = \frac{c_F}{c_1}. \tag{6.30}$$

Table 6.6 Optimum β^* and asymptotic $\tilde{\beta}$ when $F(t) = 1 - e^{-t^\alpha}$

c_F/c_1	$\alpha = 1$		$\alpha = 2$		$\alpha = 3$	
	β^*	$\tilde{\beta}$	β^*	$\tilde{\beta}$	β^*	$\tilde{\beta}$
1	1.87	2.34	1.42	1.34	1.30	1.07
2	2.71	2.98	2.01	1.68	1.80	1.31
5	4.48	4.56	3.12	2.51	2.72	1.89
10	6.72	6.74	4.38	3.65	3.69	2.70
20	10.41	10.41	6.35	5.57	5.12	4.06
50	19.59	19.59	11.20	10.35	8.57	7.42
100	32.65	32.65	18.12	17.13	13.61	12.16

Furthermore, from Example 6.2, the asymptotic expected cost rate is

$$\tilde{C}_2(\beta) = \frac{c_1\beta + c_F}{(\log \beta + \gamma)^{1/\alpha}}, \tag{6.31}$$

and an optimum $\tilde{\beta}$ which minimizes $\tilde{C}_2(\beta)$ satisfies

$$\beta[\alpha(\log \beta + \gamma) - 1] = \frac{c_F}{c_1}, \tag{6.32}$$

which agrees with (6.28) when $\alpha = 1$.

Table 6.6 presents optimum β^* in (6.30) and asymptotic $\tilde{\beta}$ in (6.32) for $\alpha = 1, 2, 3$ when $F(t) = 1 - \exp(-t^\alpha)$. Tables 6.5 and 6.6 show that all values of $n^*, \tilde{n}, \beta^*$ and $\tilde{\beta}$ increase with c_F/c_1 and decrease with α, and differences among them are small. In particular, $\beta^* < \tilde{\beta}$, and its differences are much small, especially for large c_F/c_1.

In addition, these tables indicate that $n^* < \beta^*$ because we have to prepare more units when the number of units is uncertain. When $\alpha = 1$, $n^* \leq \tilde{n}$ and $\beta^* \leq \tilde{\beta}$, and $\tilde{\beta}$ is less than c_F/c_1 for $c_F/c_1 \geq 5$, as shown in (6.28). Optimum n^* and β^* are almost the same as $\tilde{\beta}$ for large c_F/c_1. Thus, it would be sufficient in practical fields to use $\tilde{\beta}$ in (6.28) as estimated values of optimum n^* and β^*. These estimations indicate also for $\alpha > 1$ that $n^* \approx \tilde{n}$, and $n^* \approx \tilde{\beta}$, and especially, that $\tilde{\beta}$ in (6.32) is very simple and could be used as a good approximation of n^* for Weibull failure times (Problem 4 in Sect. 6.5). □

6.2.2 Optimum Replacement Time

Suppose that the system is replaced at time T ($0 < T \leq \infty$) or at failure, whichever occurs first, Then, the mean time to replacement is

$$\mu_n(T) \equiv \int_0^T \left[1 - F(t)^n\right] dt, \tag{6.33}$$

where note that $\mu_n(\infty) = \mu_n$ in (6.1). Thus, the expected cost rate in (6.16) is [8, p. 10]

$$C_1(T) = \frac{c_1 n + c_F F(T)^n}{\mu_n(T)}. \tag{6.34}$$

When $n = 1$, $C_1(T)$ agrees with the expected cost rate for the standard age replacement in (2.2), replacing c_T with c_1 and $c_F - c_T$ with c_F.

We find an optimum replacement time T_1^* which minimizes $C_1(T)$ for a given n $(n \geq 2)$. Then, differentiating $C_1(T)$ with respect to T and setting it equal to zero,

$$H_n(T)\mu_n(T) - F(T)^n = \frac{c_1 n}{c_F}, \tag{6.35}$$

where

$$H_n(t) \equiv \frac{n h(t)[F(t)^{n-1} - F(t)^n]}{1 - F(t)^n}.$$

It is easily proved that

$$\frac{1 - F(t)^{n-1}}{1 - F(t)^n} = \frac{\sum_{j=0}^{n-2} F(t)^j}{\sum_{j=0}^{n-1} F(t)^j}$$

decreases strictly with t from 1 to $(n-1)/n$ for $n \geq 2$ and

$$\lim_{t \to \infty} \frac{n[F(t)^{n-1} - F(t)^n]}{1 - F(t)^n} = 1.$$

Thus, $H_n(t)$ increases strictly with t to $h(\infty)$ for $n \geq 2$ (Problem 5 in Sect. 6.5). Denoting the left-hand side of (6.35) by $L_1(T)$, it follows that

$$\lim_{T \to 0} L_1(T) = 0, \quad \lim_{T \to \infty} L_1(T) = \mu_n h(\infty) - 1,$$
$$L_1'(T) = H_n'(T)\mu_n(T) > 0.$$

Therefore, we have the following optimum policy:

(i) If $\mu_n h(\infty) > (c_1 n + c_F)/c_F$, then there exists a finite and unique T_1^* $(0 < T_1^* < \infty)$ which satisfies (6.35), and the resulting cost rate is

$$C_1(T_1^*) = c_F H_n(T_1^*). \tag{6.36}$$

(ii) If $\mu_n h(\infty) \leq (c_1 n + c_F)/c_F$, then $T_1^* = \infty$, i.e., the system is replaced only at failure, and the expected cost rate $C_1(\infty)$ is given in (6.16).

In particular, when $F(t) = 1 - e^{-\lambda t}$, if $\sum_{j=2}^{n}(1/j) > c_1 n/c_F$ for $n \geq 2$, then T_1^* $(0 < T_1^* < \infty)$ is given by a finite and unique solution of the equation

$$\frac{ne^{-\lambda T}(1 - e^{-\lambda T})^{n-1}}{1 - (1 - e^{-\lambda T})^n} \sum_{j=1}^{n} \frac{1}{j}(1 - e^{-\lambda T})^j - (1 - e^{-\lambda T})^n = \frac{c_1 n}{c_F}, \tag{6.37}$$

and the resulting cost rate is

$$\frac{C_1(T_1^*)}{\lambda} = \frac{c_F n e^{-\lambda T_1^*}(1 - e^{-\lambda T_1^*})^{n-1}}{1 - (1 - e^{-\lambda T_1^*})^n}. \tag{6.38}$$

When $\sum_{j=2}^{n}(1/j) \leq c_1 n/c_F$ or $n = 1$, $T_1^* = \infty$, i.e., we should make no preventive replacement.

Next, consider a parallel system with N units in which N is a random variable with a Poisson distribution in (6.13). Then, the mean time to replacement is, from (6.14) and (6.33),

$$\mu_\beta(T) = \int_0^T [1 - e^{-\beta \overline{F}(t)}] \, dt, \tag{6.39}$$

and from (6.34),

$$C_2(T) = \frac{c_1 \beta + c_F e^{-\beta \overline{F}(T)}}{\int_0^T [1 - e^{-\beta \overline{F}(t)}] \, dt}, \tag{6.40}$$

which agrees with $C_2(\beta)$ in (6.25) when $T = \infty$.

In particular, when $F(t) = 1 - e^{-\lambda t}$, $C_2(T)$ becomes

$$C_2(T) = \frac{c_1 \beta + c_F \exp(-\beta e^{-\lambda T})}{\int_0^T [1 - \exp(-\beta e^{-\lambda t})] \, dt}. \tag{6.41}$$

We find an optimum time T_2^* which minimizes $C_2(T)$. Differentiating $C_2(T)$ with respect to T and setting it equal to zero,

$$H_\beta(T) \int_0^T [1 - \exp(-\beta e^{-\lambda t})] \, dt - \exp(-\beta e^{-\lambda T}) = \frac{c_1 \beta}{c_F}, \tag{6.42}$$

where

$$H_\beta(T) \equiv \frac{\lambda \beta \exp(-\lambda T - \beta e^{-\lambda T})}{1 - \exp(-\beta e^{-\lambda T})}.$$

Clearly,

$$H_\beta(0) \equiv \lim_{T \to 0} H_\beta(T) = \frac{\lambda \beta e^{-\beta}}{1 - e^{-\beta}},$$

$$H_\beta(\infty) \equiv \lim_{T \to \infty} H_\beta(T) = \lambda > H_\beta(0),$$

$$H_\beta'(T) = \frac{\lambda^2 \beta \exp(-\lambda T - \beta e^{-\lambda T})}{[1 - \exp(-\beta e^{-\lambda T})]^2} [\beta e^{-\lambda T} - 1 + \exp(-\beta e^{-\lambda T})] > 0.$$

Thus, $H_\beta(T)$ increases strictly with T from $H_\beta(0)$ to λ, which implies that the left-hand side of (6.42) increases strictly from $-e^{-\beta}$ to $\log \beta - (1 - \gamma)$. Therefore, if $\log \beta + \gamma > (c_1\beta + c_F)/c_F$ for $\beta > 1.53$, then there exists a finite and unique T_2^* $(0 < T_2^* < \infty)$, which satisfies (6.42), and the resulting cost rate is

$$C_2(T_2^*) = c_F H_\beta(T_2^*).$$

Example 6.6 (Replacement for Weibull failure time) When $F(t) = 1 - \exp(-t^\alpha)$ $(\alpha \geq 1)$, the expected cost rate in (6.34) is

$$C_1(T) = \frac{c_1 n + c_F[1 - \exp(-T^\alpha)]^n}{\int_0^T \{1 - [1 - \exp(-t^\alpha)]^n\} dt}. \tag{6.43}$$

From (6.35), an optimum T_1^* to minimize $C_1(T)$ is given by a finite and unique solution of the following equation when $\sum_{j=2}^n (1/j) > c_1 n/c_F$ for $\alpha = 1$, or for $\alpha > 1$,

$$\frac{n\alpha T^{\alpha-1} e^{-T^\alpha}(1 - e^{-T^\alpha})^{n-1}}{1 - (1 - e^{-T^\alpha})^n} \int_0^T \{1 - [1 - \exp(-t^\alpha)]^n\} dt$$

$$- (1 - e^{-T^\alpha})^n = \frac{c_1 n}{c_F}. \tag{6.44}$$

Next, when N has a Poisson distribution in (6.13), the expected cost rate in (6.40) is

$$C_2(T) = \frac{c_1 \beta + c_F \exp(-\beta e^{-T^\alpha})}{\int_0^T [1 - \exp(-\beta e^{-t^\alpha})] dt}, \tag{6.45}$$

and an optimum T_2^* to minimize $C_2(T)$ is given by

$$\frac{\beta \alpha T^{\alpha-1} \exp(-T^\alpha - \beta e^{-T^\alpha})}{1 - \exp(-\beta e^{-T^\alpha})} \int_0^T [1 - \exp(-\beta e^{-t^\alpha})] dt - \exp(-\beta e^{-T^\alpha}) = \frac{\beta c_1}{c_F}. \tag{6.46}$$

Table 6.7 Optimum T_1^* and T_2^* when $F(t) = 1 - e^{-t^\alpha}$ and $n = \beta = 10$

c_F/c_1	$\alpha = 1$		$\alpha = 2$		$\alpha = 3$	
	T_1^*	T_2^*	T_1^*	T_2^*	T_1^*	T_2^*
1	∞	∞	3.31	2.53	1.66	1.51
2	∞	∞	1.96	1.72	1.39	1.30
5	∞	∞	1.48	1.39	1.21	1.18
10	2.44	2.11	1.30	1.21	1.15	1.09
20	1.78	1.57	1.18	1.09	1.09	1.03
50	1.36	1.15	1.06	0.97	1.00	0.94
100	1.15	0.94	1.00	0.88	0.97	0.87

When $\alpha = 1$, (6.44) and (6.46) becomes (6.37) and (6.42) for $\lambda = 1$, respectively.

Table 6.7 presents optimum T_1^* and T_2^* which satisfy (6.44) and (6.46) for $\alpha > 1$, and (6.37) and (6.42) for $\alpha = 1$, respectively, when $\beta = n = 10$. Both optimum T_1^* and T_2^* decrease with c_F/c_1 and α. This indicates that $T_1^* > T_2^*$, because if the number of units is uncertain, then we should replace the system earlier than that with constant n. However, the differences between T_1^* and T_2^* are small as c_F/c_1 and α increase. When $\alpha = 1$, if $(1/10) \sum_{j=2}^{10}(1/j) = 0.192 \le c_1/c_F$, then $T_1^* = \infty$, and if $(1/10)[\log 10 - (1 - \gamma)] = 0.188 \le c_1/c_F$, then $T_2^* = \infty$. $\square$

6.3 K-out-of-n System

Consider a K-out-of-n system ($K = 1, 2, \ldots, n$) in which it is operating if and only if at least K units are operating [1, p. 216], [8, p. 12]. Then, we try to rewrite the results in Sects. 6.1 and 6.2 for a K-out-of-n system when K is constant and random.

6.3.1 Constant K

Suppose that K is constant. The reliability of the system at time t is [1, p. 216], [8, p. 12]

$$R(t) = \sum_{j=0}^{K-1} \binom{n}{j} [F(t)]^j [\overline{F}(t)]^{n-j}$$

$$= \sum_{j=K}^{n} \binom{n}{j} [\overline{F}(t)]^j [F(t)]^{n-j} \quad (K = 1, 2, \ldots, n), \tag{6.47}$$

which decreases with t from 0 to 1, because

$$R'(t) = nf(t) \sum_{j=K}^{n} \left\{ \binom{n-1}{j} [\overline{F}(t)]^j [F(t)]^{n-j-1} - \binom{n-1}{j-1} [\overline{F}(t)]^{j-1} [F(t)]^{n-j} \right\}$$

$$= -nf(t) \binom{n-1}{K-1} [\overline{F}(t)]^{K-1} [F(t)]^{n-K} \le 0.$$

The MTTF is

$$\mu_{n,K} = \int_0^\infty R(t)\, dt = \sum_{j=0}^{K-1} \binom{n}{j} \int_0^\infty [F(t)]^j [\overline{F}(t)]^{n-j}\, dt$$

$$= \sum_{j=K}^{n} \binom{n}{j} \int_0^\infty [\overline{F}(t)]^j [F(t)]^{n-j}\, dt \quad (K = 1, 2, \ldots, n), \quad (6.48)$$

which decreases with K from μ_n in (6.1) to $\int_0^\infty [\overline{F}(t)]^n\, dt$. In particular, when $F(t) = 1 - e^{-\lambda t}$,

$$\mu_{n,K} = \frac{1}{\lambda} \sum_{j=K}^{n} \frac{1}{j}, \qquad (6.49)$$

and using (6.3), it is approximately given by

$$\tilde{\mu}_{n,K} = \frac{1}{\lambda}[\log n - \log(K-1)] = \frac{1}{\lambda} \log \frac{n}{K-1} \qquad (6.50)$$

for $K \ge 2$. In addition, when $F(t) = 1 - \exp[-(\lambda t)^\alpha]$ ($\alpha \ge 1$), using (6.6) and (6.50),

$$\tilde{\mu}_{n,K} = \frac{1}{\lambda} \left(\sum_{j=K}^{n} \frac{1}{j} \right)^{1/\alpha} \approx \frac{1}{\lambda} \left(\log \frac{n}{K-1} \right)^{1/\alpha}. \qquad (6.51)$$

Example 6.7 (MTTF for Weibull failure time) Table 6.8 presents exact MTTF, approximations $[\sum_{j=K}^{n} (1/j)]^{1/2}$ and $\log[n/(K-1)]^{1/2}$ when $F(t) = 1 - \exp(-t^2)$ and $n = 100$. These approximations give good ones to exact MTTF for any K because $n = 100$ is large, and $\mu_{n,K} < [\sum(1/j)]^{1/2} \le [\log n/(K-1)]^{1/2}$. When $K = 1$, MTTF is given in Table 6.2.   □

Next, from (6.16), the expected cost rate is [8, p. 12]

$$C_1(n, K) = \frac{c_1 n + c_F}{\mu_{n,K}} \qquad (n = K, K+1, \ldots). \qquad (6.52)$$

Table 6.8 Exact MTTF and its approximations when $F(t) = 1 - e^{-t^2}$ and $n = 100$

K	$\mu_{n,K}$	$[\sum(1/j)]^{1/2}$	$[\log n/(K-1)]^{1/2}$
2	2.037	2.046	2.146
5	1.757	1.762	1.794
10	1.533	1.536	1.552
20	1.278	1.281	1.289
50	0.839	0.842	0.845
60	0.722	0.724	0.726
70	0.605	0.607	0.609
80	0.481	0.484	0.486
90	0.337	0.341	0.341
100	0.089	0.100	0.100

In particular, when $F(t) = 1 - e^{-\lambda t}$, the expected cost rate is, from (6.49),

$$\frac{C_1(n, K)}{\lambda} = \frac{c_1 n + c_F}{\sum_{j=K}^{n}(1/j)}. \tag{6.53}$$

From the inequality $C_1(n + 1, K) - C_1(n, K) \geq 0$,

$$(n+1) \sum_{j=K}^{n} \frac{1}{j} - n \geq \frac{c_F}{c_1}, \tag{6.54}$$

whose left-hand side increases strictly with n to ∞. Thus, there exists a finite and unique minimum n^* ($K \leq n^* < \infty$) which satisfies (6.54) and increases with K.

Furthermore, suppose that the system is replaced at time T ($0 < T \leq \infty$) or at failure, whichever occurs first. Then, the expected cost is, from (6.47) and (6.48)

$$C_1(T; K) = \frac{c_1 n + c_F \sum_{j=0}^{K-1} \binom{n}{j}[\overline{F}(T)]^j [F(T)]^{n-j}}{\sum_{j=K}^{n} \binom{n}{j} \int_0^T [\overline{F}(t)]^j [F(t)]^{n-j} dt}, \tag{6.55}$$

where c_1 and c_F are given in (6.34). Differentiating $C_1(T; K)$ with respect to T and setting it equal to zero,

$$H_n(T; K) \sum_{j=K}^{n} \binom{n}{j} \int_0^T [\overline{F}(t)]^j [F(t)]^{n-j} dt - \sum_{j=0}^{K-1} \binom{n}{j}[\overline{F}(T)]^j [F(T)]^{n-j} = \frac{c_1 n}{c_F}, \tag{6.56}$$

where

$$H_n(T; K) \equiv \frac{n h(T) \binom{n-1}{K-1}[\overline{F}(T)]^K [F(T)]^{n-K}}{\sum_{j=K}^{n} \binom{n}{j}[\overline{F}(T)]^j [F(T)]^{n-j}}.$$

Note that if $H_n(T; K)$ increases strictly with T, then the left-hand side of (6.56) also increases strictly with T. In particular, when $F(t) = 1 - e^{-\lambda t}$ for $n \geq 2$,

$$H_n(T; K) = \frac{n\lambda \binom{n-1}{K-1}}{\sum_{j=K}^{n} \binom{n}{j}(e^{\lambda T} - 1)^{K-j}}, \tag{6.57}$$

which increases with T from 0 to $K\lambda$ for $K < n$ and is constant $n\lambda$ for $K = n$. Therefore, if $K < n$ and

$$\frac{K}{n} \sum_{j=K+1}^{n} \frac{1}{j} > \frac{c_1}{c_F},$$

then there exists a finite and unique T_1^* ($0 < T_1^* < \infty$) which satisfies (6.56) (Problem 6 in Sect. 6.5). When $K = n$, any finite T_1^* does not exist, i.e., $T_1^* = \infty$. Furthermore, when $n = 2$ and $K = 1$, if $c_F \leq 4c_1$ then $T_1^* = \infty$.

6.3.2 Random K

It is assumed that K is a random variable for a specified n ($n \geq 1$), i.e., $p_{k;n} \equiv \Pr\{K = k\}$ ($k = 1, 2, \ldots, n$) and $\sum_{k=1}^{n} p_{k;n} = 1$. Then, the reliability at time t is, from (6.47),

$$R(t; p) = \sum_{k=1}^{n} p_{k;n} \sum_{j=0}^{k-1} \binom{n}{j}[F(t)]^j[\overline{F}(t)]^{n-j}$$

$$= \sum_{j=1}^{n} \binom{n}{j}[\overline{F}(t)]^j[F(t)]^{n-j} \sum_{k=1}^{j} p_{k;n}, \tag{6.58}$$

and MTTF is

$$\mu_{n,p} = \sum_{j=1}^{n} \binom{n}{j} \int_0^\infty [\overline{F}(t)]^j[F(t)]^{n-j} \, dt \sum_{k=1}^{j} p_{k;n}. \tag{6.59}$$

In particular, when $F(t) = 1 - e^{-\lambda t}$,

$$\mu_{n,p} = \frac{1}{\lambda} \sum_{j=1}^{n} \frac{1}{j} \sum_{k=1}^{j} p_{k;n} = \frac{1}{\lambda} \sum_{k=1}^{n} p_{k;n} \sum_{j=k}^{n} \frac{1}{j}. \tag{6.60}$$

From (6.50), it is approximately,

$$\tilde{\mu}_{n,p} = \frac{1}{\lambda} \sum_{k=1}^{n} p_{k;n} \log \left(\frac{n}{k-1} \right) \tag{6.61}$$

for large n, where when $k = 1$, $\log[n/(k-1)] \equiv \log n$.

Example 6.8 (MTTF for Poisson distribution) When $p_{k;n} = [\beta^{k-1}/(k-1)!]/ \sum_{j=0}^{n-1}(\beta^j/j!)$ $(k = 1, 2, \ldots, n; 0 < \beta < \infty)$ with mean

$$E\{K\} = \sum_{k=1}^{n} k p_{k;n} = \frac{\beta \sum_{j=0}^{n-2}(\beta^j/j!)}{\sum_{j=0}^{n-1}(\beta^j/j!)} + 1,$$

where $\sum_{j=0}^{-1} \equiv 0$, MTTF in (6.60) is

$$\mu_{n,p} = \frac{\sum_{j=1}^{n}(1/j) \sum_{k=0}^{j-1}(\beta^k/k!)}{\lambda \sum_{k=0}^{n-1}(\beta^k/k!)}, \tag{6.62}$$

which decreases with β from $\mu_{n,K}$ in (6.49) to $1/(n\lambda)$ (Problem 7 in Sect. 6.5).
For large n, i.e., as $\sum_{j=0}^{n-1}(\beta^j/j!)e^{-\beta} \to 1$,

$$\tilde{\mu}_{n,p} = \frac{1}{\lambda} \sum_{j=1}^{n} \frac{1}{j} \sum_{k=0}^{j-1} \frac{\beta^k}{k!} e^{-\beta}. \tag{6.63}$$

Table 6.9 presents $\mu_{n,p}$ and $\tilde{\mu}_{n,p}$ when $n = 100$ and $E\{K\} = k$ (Problem 8 in Sect. 6.5). Because $n = 100$ is large, $\tilde{\mu}_{n,p}$ gives a good approximation to the exact $\mu_{n,p}$ except large k. Compared to Table 6.8, it is of interest that $\mu_{n,p}$ decreases with k from 5.187 to $1/n = 0.01$. □

Next, when $F(t) = 1 - e^{-\lambda t}$ and $p_{k;n} = \Pr\{K = k\}$ $(k = 1, 2, \ldots, n)$, the expected cost rate is, from (6.52) and (6.60),

$$\frac{C_2(n; p)}{\lambda} = \frac{c_1 n + c_F}{\mu_{n,p}} \quad (n = 1, 2, \ldots). \tag{6.64}$$

Example 6.9 (Number of units for Poisson distribution) When $p_{k;n} = [\beta^{k-1}/(k-1)!]/ \sum_{j=0}^{n-1}(\beta^j/j!)$ $(k = 1, 2, \ldots, n)$, $\mu_{n,p}$ is given in (6.62) and the expected cost rate in (6.64) is

Table 6.9 MTTF $\mu_{n,p}$ and approximation $\tilde{\mu}_{n,p}$ when $F(t) = 1 - e^{-t}$, $n = 100$ and $E\{K\} = k$

k	$\mu_{n,p}$	$\tilde{\mu}_{n,p}$
1	5.187	5.187
2	4.391	4.391
5	3.220	3.220
10	2.413	2.413
20	1.666	1.666
50	0.718	0.718
60	0.533	0.533
70	0.376	0.376
80	0.240	0.241
90	0.119	0.127
100	0.011	0.048

$$\frac{C_2(n; p)}{\lambda} = \frac{(c_1 n + c_F) \sum_{k=0}^{n-1} (\beta^k / k!)}{\sum_{j=1}^{n} (1/j) \sum_{k=0}^{j-1} (\beta^k / k!)}. \tag{6.65}$$

When n is large, i.e., $\sum_{k=0}^{n-1} (\beta^k / k!) e^{-\beta} \approx 1$, MTTF is given in (6.63), and the asymptotic expected cost rate is

$$\frac{\tilde{C}(n; p)}{\lambda} = \frac{c_1 n + c_F}{\displaystyle\sum_{j=1}^{n} (1/j) \sum_{k=0}^{j-1} (\beta^k / k!) e^{-\beta}}. \tag{6.66}$$

From the inequality $\tilde{C}(n + 1; p) - \tilde{C}(n; p) \geq 0$,

$$(n + 1) \sum_{j=1}^{n} \frac{1}{j} \sum_{k=0}^{j-1} \frac{\beta^k}{k!} e^{-\beta} - n \geq \frac{c_F}{c_1}, \tag{6.67}$$

whose left-hand side increases strictly with n from $2e^{-\beta} - 1$ to ∞. Thus, there exists a finite and unique $\tilde{n}_p$ $(1 \leq \tilde{n}_p < \infty)$ which satisfies (6.67). Note that $2e^{-\beta} - 1 < 0$ for $\beta \geq 1$.

Table 6.10 presents optimum n^* which satisfies (6.54), n_p^* which minimizes $C_2(n; p)$ in (6.65) and $\tilde{n}_p$ which satisfies (6.67) when $E\{K\} = k$. This indicates that all of n^*, n_p^* and $\tilde{n}_p$ increase with k and c_F/c_1, and are almost the same. This also shows that n^*/k decreases with k, and in this case, $n^* \geq n_p^* = \tilde{n}_p$ for all k (Problem 9 in Sect. 6.5). □

Furthermore, when the system is replaced at time T, the expected cost rate is, from (6.55) and (6.59),

Table 6.10 Optimum n^*, n_p^* and $\widetilde{n}_p$ when $F(t) = 1 - e^{-t}$ and $E\{K\} = k$

k	$c_F/c_1 = 50$			$c_F/c_1 = 100$		
	n^*	n_p^*	$\widetilde{n}_p$	n^*	n_p^*	$\widetilde{n}_p$
1	19	19	19	32	32	32
2	26	24	24	42	40	40
5	40	38	38	61	58	58
7	48	46	46	71	69	69
10	59	57	57	84	82	82
20	91	89	89	120	119	119
30	120	119	119	153	151	151
40	149	148	148	184	182	182

$$C_2(T; p) = \frac{c_1 n + c_F \sum_{k=1}^{n} p_{k;n} \sum_{j=0}^{k-1} \binom{n}{j} [\overline{F}(T)]^j [F(T)]^{n-j}}{\sum_{k=1}^{n} p_{k;n} \sum_{j=k}^{n} \binom{n}{j} \int_0^T [\overline{F}(t)]^j [F(t)]^{n-j} \, dt}. \tag{6.68}$$

Differentiating $C_2(T; p)$ with respect to T and setting it equal to zero,

$$H_n(T; p) \sum_{k=1}^{n} p_{k;n} \sum_{j=k}^{n} \binom{n}{j} \int_0^T [\overline{F}(t)]^j [F(t)]^{n-j} \, dt$$

$$- \sum_{k=1}^{n} p_{k;n} \sum_{j=0}^{k-1} \binom{n}{j} [\overline{F}(T)]^j [F(T)]^{n-j} = \frac{c_1 n}{c_F}, \tag{6.69}$$

whose left-hand side increases strictly with T if $H_n(T; p)$ increases strictly, where

$$H_n(T; p) \equiv \frac{nh(T) \sum_{k=1}^{n} p_{k;n} \binom{n-1}{k-1} [\overline{F}(T)]^k [F(T)]^{n-k}}{\sum_{k=1}^{n} p_{k;n} \sum_{j=k}^{n} \binom{n}{j} [\overline{F}(T)]^j [F(T)]^{n-j}}.$$

Example 6.10 (Replacement for Poisson distribution) When $F(t) = 1 - e^{-\lambda t}$ and $p_{k;n} = [\beta^{k-1}/(k-1)!] / \sum_{j=0}^{n-1} (\beta^j/j!)$ for $n \geq 2$, (6.69) becomes

$$H_n(T; p) \sum_{k=0}^{n-1} \frac{\beta^k}{k!} \sum_{j=k+1}^{n} \int_0^T B_j(t) \, dt - \sum_{k=0}^{n-1} \frac{\beta^k}{k!} \sum_{j=0}^{k} B_j(T) = \frac{c_1 n}{c_F} \sum_{k=0}^{n-1} \frac{\beta^k}{k!}, \tag{6.70}$$

where

$$B_j(T) \equiv \binom{n}{j} (e^{-\lambda T})^j (1 - e^{-\lambda T})^{n-j} \quad (j = 0, 1, \ldots, n),$$

$$H_n(T; p) = \frac{\lambda \sum_{k=0}^{n-1} [(k+1)\beta^k/k!] B_{k+1}(T)}{\sum_{k=0}^{n-1} (\beta^k/k!) \sum_{j=k+1}^{n} B_j(T)}.$$

Table 6.11 Optimum T^*, T_p^*, and $\tilde{T}_p$ when $F(t) = 1 - e^{-t}$, $n = 100$ and $E\{K\} = k$

k	$c_F/c_1 = 50$			$c_F/c_1 = 100$		
	T^*	T_p^*	$\tilde{T}_p$	T^*	T_p^*	$\tilde{T}_p$
1	4.48	4.48	4.50	3.87	3.87	3.87
2	3.59	3.78	3.80	3.25	3.23	3.24
5	2.68	2.74	2.76	2.49	2.44	2.44
7	2.37	2.41	2.43	2.22	2.16	2.16
10	2.05	2.08	2.09	1.93	1.87	1.87
20	1.43	1.45	1.47	1.35	1.30	1.31
30	1.07	1.09	1.12	1.01	0.97	0.98
40	0.81	0.85	0.89	0.76	0.74	0.75

Letting $L_2(T)$ be the left-hand side of (6.70),

$$L_2(0) \equiv \lim_{T \to 0} L_2(T) = 0,$$

$$L_2(\infty) \equiv \lim_{T \to \infty} L_2(T) = \sum_{k=0}^{n-1} \frac{\beta^k}{k!} \left(\sum_{j=k+1}^{n} \frac{1}{j} - 1 \right),$$

$$L_2'(T) = H_n'(T; p) \sum_{k=0}^{n-1} \frac{\beta^k}{k!} \sum_{j=k+1}^{n} \int_0^T B_j(t)\, dt.$$

Thus, if $H_n(T; p)$ increases strictly with T and $L_2(\infty) > (c_1 n/c_F) \sum_{k=0}^{n-1}(\beta^k/k!)$ then there exists an optimum T_p^* ($0 < T_p^* < \infty$) which satisfies (6.70). Furthermore, when n is large, i.e., $\sum_{k=0}^{n-1}(\beta^k/k!)e^{-\beta} \approx 1$, asymptotic $\tilde{T}_p$ satisfies

$$H_n(T; p) \sum_{k=0}^{n-1} \frac{\beta^k}{k!} e^{-\beta} \sum_{j=k+1}^{n} \int_0^T B_j(t)\, dt - \sum_{k=0}^{n-1} \frac{\beta^k}{k!} e^{-\beta} \sum_{j=0}^{k} B_j(T) = \frac{c_1 n}{c_F}, \quad (6.71)$$

and $\tilde{T}_p \geq T_p^*$.

Table 6.11 presents optimum T^* which satisfies (6.56), T_p^* which satisfies (6.70) and $\tilde{T}_p$ which satisfies (6.71) when $n = 100$ and $E\{K\} = k$. This indicates that all of T^*, T_p^* and $\tilde{T}_p$ decrease with k and c_F/c_1, and are almost the same. This also shows that $T_p^* \leq \tilde{T}_p$.

6.4 Shortage and Excess Costs

Introducing shortage and excess costs, we will take up the problems how to determine the scheduling time, and what kinds of redundant systems to provide for a job, which will be discussed in Chap. 7. Using such shortage and excess costs, it has been shown in Chap. 2 that even when the failure time is exponential, a finite replacement time exists.

This section attempts to introduce the shortage and excess costs to a parallel system discussed in Sect. 6.2: Optimum replacement time and random replacement time which minimize the expected costs are derived analytically. Conversely, when the replacement time is fixed, an optimum number of units is obtained. Furthermore, when a parallel system for a job with random working times is replaced at Nth working time, an optimum number N^* is given. Finally, when the number of units is random, two expected costs are obtained, and optimum replacement times which minimize them are derived. In addition, when the random number of units has a Poisson distribution, optimum times are computed numerically.

6.4.1 Age Replacement

Suppose that a parallel system with n units is replaced before failure at time T ($0 < T \leq \infty$) as the preventive replacement. Then, introduce the two kinds of linear costs in Fig. 2.1 which depend only on time length: If the system fails after time T, then this causes a shortage cost $c_S(X - T)$ because the system might operate for a little more time. On the other hand, if the system would fail before time T, then this causes an excess cost $c_E(T - X)$ due to its failure because it fails at a little earlier than time T and the replacement was estimated to be planned longer than an actual failure time.

Under the above conditions, the expected replacement cost is, from (2.6),

$$
C_1(T; n) = c_S \int_T^\infty (t - T) \, dF(t)^n + c_E \int_0^T (T - t) \, dF(t)^n
$$

$$
= c_S \int_T^\infty [1 - F(t)^n] \, dt + c_E \int_0^T F(t)^n \, dt. \tag{6.72}
$$

Thus, because the mean time to replacement is given in (6.33), the expected cost rate is

$$
C_2(T; n) = \frac{c_S \int_T^\infty [1 - F(t)^n] \, dt + c_E \int_0^T F(t)^n \, dt}{\int_0^T [1 - F(t)^n] \, dt}. \tag{6.73}
$$

We find analytically optimum T_i^* ($i = 1, 2$) which minimize $C_i(T; n)$ for a given n ($n \geq 1$). Clearly,

$$\lim_{T \to 0} C_1(T; n) = c_S \mu_n, \qquad \lim_{T \to \infty} C_1(T; n) = \infty,$$

where μ_n is given in (6.1). Thus, there exists an optimum T_1^* ($0 \le T_1^* < \infty$) which minimizes $C_1(T; n)$ in (6.72). Differentiating $C_1(T; n)$ with respect to T and setting it equal to zero,

$$F(T)^n = \frac{c_S}{c_S + c_E}. \tag{6.74}$$

Next, from (6.73),

$$\lim_{T \to 0} C_2(T; n) = \lim_{T \to \infty} C_2(T; n) = \infty.$$

Thus, there exits an optimum T_2^* ($0 < T_2^* < \infty$) which minimizes $C_2(T; n)$. Differentiating $C_2(T; n)$ with respect to T and setting it equal to zero,

$$\frac{\int_0^T [1 - F(t)^n] \, dt}{1 - F(T)^n} - T = \frac{c_S \mu_n}{c_E}, \tag{6.75}$$

whose left-hand side increases strictly from 0 to ∞. Therefore, there exists a finite and unique T_2^* ($0 < T_2^* < \infty$) which satisfies (6.75), and the resulting cost rate is

$$C_2(T_2^*; n) = \frac{c_E}{1 - F(T_2^*)^n} - (c_S + c_E). \tag{6.76}$$

It can be seen that $T_2^* \ge T_1^*$ (Problem 10 in Sect. 6.5).

In particular, when $F(t) = 1 - e^{-\lambda t}$, from (6.74), T_1^* is given by

$$1 - e^{-\lambda T} = \left(\frac{c_S}{c_S + c_E} \right)^{1/n},$$

and T_2^* satisfies uniquely

$$\frac{\int_0^T [1 - (1 - e^{-\lambda t})^n] \, dt}{1 - (1 - e^{-\lambda T})^n} - T = \frac{c_S}{c_E \lambda} \sum_{j=1}^{n} \frac{1}{j}.$$

or

$$\frac{\sum_{j=1}^n (1 - e^{-\lambda T})^j / j}{1 - (1 - e^{-\lambda T})^n} - \lambda T = \frac{c_S}{c_E} \sum_{j=1}^{n} \frac{1}{j}. \tag{6.77}$$

It is shown that T_2^* increases strictly with n: Denoting

$$L_2(T) \equiv \sum_{j=1}^{n+1} \frac{(1 - e^{-\lambda T})^{j-1}}{j} - \sum_{j=1}^{n} \frac{(1 - e^{-\lambda T})^j}{j} - \frac{1}{n+1},$$

we have

$$\lim_{T \to 0} L_2(T) = 1 - \frac{1}{n+1} > 0, \quad \lim_{T \to \infty} L_2(T) = 0,$$

$$L_2'(T) = -\lambda e^{-\lambda T} \sum_{j=1}^{n} \frac{(1 - e^{-\lambda T})^{j-1}}{j+1} < 0,$$

which implies that $L_2(T)$ decreases from $n/(n+1)$ to 0, i.e., $L_2(T) \geq 0$. So that, the left-hand side of (6.77) decreases strictly with n, and hence, an optimum T_2^* increases strictly with n because the right-hand increases with n.

6.4.2 Random Replacement

Suppose that a planned time T is not constant and is a random variable with a general distribution $G(t) \equiv \Pr\{T \leq t\}$ with finite mean $1/\theta \equiv \int_0^\infty \overline{G}(t)\, dt$ $(0 < \theta < \infty)$. The other notations and assumptions are the same ones in Sect. 6.4.1. Then, the expected replacement cost is, from (6.72),

$$
\begin{aligned}
C_1(G; n) =& c_S \int_0^\infty \left[\int_0^t (t - u)\, dG(u) \right] dF(t)^n \\
&+ c_E \int_0^\infty \left[\int_t^\infty (u - t)\, dG(u) \right] dF(t)^n \\
=& c_S \int_0^\infty [1 - F(t)^n] G(t)\, dt + c_E \int_0^\infty F(t)^n \overline{G}(t)\, dt.
\end{aligned}
\tag{6.78}
$$

Furthermore, because the mean time to replacement is, from (6.33),

$$\int_0^\infty \left\{ \int_0^u [1 - F(t)^n]\, dt \right\} dG(u) = \int_0^\infty [1 - F(t)^n] \overline{G}(t)\, dt,$$

the expected cost rate is

$$C_2(G; n) = \frac{c_S \int_0^\infty [1 - F(t)^n] G(t)\, dt + c_E \int_0^\infty F(t)^n \overline{G}(t)\, dt}{\int_0^\infty [1 - F(t)^n] \overline{G}(t)\, dt}. \tag{6.79}$$

In particular, when $G(t) = 1 - e^{-\theta t}$, the expected costs are the function of θ, and from (6.78),

$$C_1(\theta; n) = c_S \int_0^\infty [1 - F(t)^n]\, dt + \frac{c_E}{\theta} - (c_S + c_E) \int_0^\infty [1 - F(t)^n] e^{-\theta t}\, dt, \tag{6.80}$$

and from (6.79),

$$C_2(\theta; n) = \frac{c_S \int_0^\infty [1 - F(t)^n]\, dt + c_E/\theta}{\int_0^\infty [1 - F(t)^n] e^{-\theta t}\, dt} - (c_S + c_E). \tag{6.81}$$

We find optimum θ_i^* ($i = 1, 2$) which minimize $C_i(\theta; n)$. From (6.80),

$$\lim_{\theta \to 0} C_1(\theta; n) = \infty, \quad \lim_{\theta \to \infty} C_1(\theta; n) = c_S \mu_n.$$

Differentiating $C_1(\theta; n)$ with respect to θ and setting it equal to zero,

$$\int_0^\infty [1 - F(t)^n] \theta^2 t e^{-\theta t}\, dt = \frac{c_E}{c_S + c_E},$$

i.e.,

$$\int_0^\infty [1 - (1 + \theta t) e^{-\theta t}]\, dF(t)^n = \frac{c_E}{c_S + c_E}, \tag{6.82}$$

whose left-hand increases with θ from 0 to 1. Therefore, there exist a finite and unique θ_1^* ($0 < \theta_1^* < \infty$) which satisfies (6.82), and $1/\theta_1^*$ increases strictly with n to ∞. Furthermore, from (6.81),

$$\lim_{\theta \to 0} C_2(\theta; n) = \lim_{\theta \to \infty} C_2(\theta; n) = \infty.$$

Differentiating $C_2(\theta; n)$ with respect to θ and setting it equal to zero,

$$\frac{\int_0^\infty t e^{-\theta t}\, dF(t)^n}{\int_0^\infty [1 - (1 + \theta t) e^{-\theta t}]\, dF(t)^n} = \frac{c_S \mu_n}{c_E}, \tag{6.83}$$

whose left-hand side decreases strictly with θ from ∞ to 0. Therefore, there exists a finite and unique θ_2^* ($0 < \theta_2^* < \infty$) which satisfies (6.83) (Problem 11 in Sect. 6.5).

Example 6.11 (Replacement for exponential failure time) Suppose that $F(t) = 1 - e^{-t}$ and $G(t) = 1 - e^{-\theta t}$. Then, from (6.74), T_1^* is

$$1 - e^{-T} = \left(\frac{c_S}{c_S + c_E}\right)^{1/n},$$

and from (6.77), T_2^* is

$$\frac{\sum_{j=1}^{n}(1 - e^{-T})^j/j}{1 - (1 - e^{-T})^n} - T = \frac{c_S}{c_E}\sum_{j=1}^{n}\frac{1}{j}.$$

From (6.82), θ_1^* is

$$\int_0^\infty [1 - (1 - e^{-t})^n]\theta^2 t e^{-\theta t}\,dt = \frac{c_E}{c_S + c_E},$$

and from (6.83), θ_2^* is

$$\frac{\int_0^\infty n(1 - e^{-t})^{n-1}e^{-t}t e^{-\theta t}\,dt}{\int_0^\infty [1 - (1 - e^{-t})^n]\theta^2 t e^{-\theta t}\,dt} = \frac{c_S}{c_E}\sum_{j=1}^{n}\frac{1}{j}.$$

All optimum T_i^* and $1/\theta_i^*$ increase strictly with n.

Table 6.12 presents optimum T_i^* and $1/\theta_i^*$ ($i = 1, 2$) when $F(t) = 1 - e^{-t}$ and $G(t) = 1 - e^{-\theta t}$. All T_i^* and θ_i^* increase with n and c_S/c_E. This indicates that $T_1^* < T_2^*$ and $1/\theta_1^* < 1/\theta_2^*$, however, $T_i^* > 1/\theta_i^*$ for small c_S/c_E and $T_i^* < 1/\theta_i^*$ for large c_S/c_E. For example, when $n = 2$, $T_1^* = 1/\theta_1^*$ for $c_S/c_E = 7.850$, and $T_2^* = 1/\theta_2^*$ for $c_S/c_E = 3.814$. □

6.4.3 Random Number of Units

Suppose that the system is replaced before failure at time T and the number of units is a random variable with a Poisson distribution in (6.13). Then, the expected replacement cost is, from (6.72),

Table 6.12 Optimum T_i^* and $1/\theta_i^*$ $(i = 1, 2)$ when $F(t) = 1 - e^{-t}$ and $G(t) = 1 - e^{-\theta t}$

c_S/c_E	$n = 2$				$n = 5$				$n = 10$			
	T_1^*	T_2^*	$\frac{1}{\theta_1^*}$	$\frac{1}{\theta_2^*}$	T_1^*	T_2^*	$\frac{1}{\theta_1^*}$	$\frac{1}{\theta_2^*}$	T_1^*	T_2^*	$\frac{1}{\theta_1^*}$	$\frac{1}{\theta_2^*}$
0.1	0.359	0.710	0.164	0.497	0.965	1.275	0.386	0.798	1.546	1.808	0.573	1.053
0.2	0.525	0.914	0.253	0.674	1.200	1.525	0.527	1.045	1.808	2.080	0.752	1.357
0.5	0.861	1.267	0.460	1.023	1.623	1.944	0.838	1.528	2.263	2.527	1.144	1.951
1.0	1.228	1.609	0.736	1.412	2.044	2.335	1.235	2.067	2.703	2.941	1.639	2.615
2.0	1.695	2.020	1.176	1.962	2.552	2.792	1.857	2.826	3.226	3.420	2.410	3.550
5.0	2.440	2.667	2.143	3.052	3.330	3.492	3.205	4.329	4.014	4.144	4.077	5.405
10.0	3.068	3.224	3.299	4.276	3.970	4.080	4.810	6.026	4.658	4.746	6.054	7.490
20.0	3.726	3.827	4.979	6.012	4.634	4.705	7.133	8.415	5.325	5.381	8.916	10.443
50.0	4.620	4.673	8.375	9.446	5.533	5.570	11.808	13.172	6.226	6.254	14.685	16.281

$$C_1(T; \beta) = \sum_{n=0}^{\infty} \frac{\beta^n}{n!} e^{-\beta} \left\{ c_S \int_T^{\infty} [1 - F(t)^n] \, dt + c_E \int_0^T F(t)^n \, dt \right\}$$

$$= c_S \int_T^{\infty} [1 - e^{-\beta \overline{F}(t)}] \, dt + c_E \int_0^T e^{-\beta \overline{F}(t)} \, dt, \qquad (6.84)$$

and the expected cost rate is, from (6.73),

$$C_2(T; \beta) = \frac{\sum_{n=0}^{\infty} (\beta^n/n!) e^{-\beta} \{ c_S \int_T^{\infty} [1 - F(t)^n] \, dt + c_E \int_0^T F(t)^n \, dt \}}{\sum_{n=0}^{\infty} (\beta^n/n!) e^{-\beta} \int_0^T [1 - F(t)^n] \, dt}$$

$$= \frac{c_S \int_T^{\infty} [1 - e^{-\beta \overline{F}(t)}] \, dt + c_E \int_0^T e^{-\beta \overline{F}(t)} \, dt}{\int_0^T [1 - e^{-\beta \overline{F}(t)}] \, dt}. \qquad (6.85)$$

We find optimum T_i^* $(i = 1, 2)$ which minimize $C_i(T; \beta)$. Differentiating $C_1(T; \beta)$ with respect to T and setting it equal to zero,

$$e^{-\beta \overline{F}(T)} = \frac{c_S}{c_S + c_E}, \qquad (6.86)$$

whose left-hand side increases strictly with T from $e^{-\beta}$ to 1. Therefore, if $e^{-\beta} < c_S/(c_S + c_E)$, then there exists a finite and unique T_1^* $(0 < T_1^* < \infty)$ which satisfies (6.86), and T_1^* increases with β. Differentiating $C_2(T; \beta)$ with respect to T and setting it equal to zero,

$$\frac{\int_0^T [1 - e^{-\beta \overline{F}(t)}] \, dt}{1 - e^{-\beta \overline{F}(T)}} - T = \frac{c_S}{c_E} \int_0^{\infty} [1 - e^{-\beta \overline{F}(t)}] \, dt, \qquad (6.87)$$

Table 6.13 Optimum T_i^* ($i = 1, 2$) when $F(t) = 1 - e^{-t}$ and N is a Poisson distribution with mean β

c_S/c_E	$\beta = 2$		$\beta = 5$		$\beta = 10$	
	T_1^*	T_2^*	T_1^*	T_2^*	T_1^*	T_2^*
0.1	0.000	0.689	0.735	1.204	1.428	1.750
0.2	0.110	0.905	1.026	1.475	1.719	2.038
0.5	0.599	1.273	1.515	1.916	2.208	2.504
1.0	1.060	1.622	1.976	2.319	2.669	2.927
2.0	1.596	2.036	2.512	2.785	3.205	3.412
5.0	2.395	2.682	3.311	3.490	4.005	4.141
10.0	3.044	3.236	3.960	4.080	4.653	4.744
20.0	3.713	3.836	4.630	4.705	5.323	5.381
50.0	4.615	4.679	5.531	5.570	6.225	6.254

whose left-hand side increases strictly with T from 0 to ∞. Therefore, there exists an optimum T_2^* ($0 < T_2^* < \infty$) which satisfies (6.87), and T_2^* increases strictly with β and $T_2^* \geq T_1^*$ (Problem 12 in Sect. 6.5).

Example 6.12 (Replacement for Poisson distribution) When $F(t) = 1 - e^{-t}$, T_1^* satisfies

$$\exp(-\beta e^{-T}) = \frac{c_S}{c_S + c_E},$$

and T_2^* satisfies

$$\frac{\int_0^T [1 - \exp(-\beta e^{-t})]\, dt}{1 - \exp(-\beta e^{-T})} - T = \frac{c_S}{c_E} \int_0^\infty [1 - \exp(-\beta e^{-t})]\, dt.$$

Table 6.13 presents optimum T_i^* when $F(t) = 1 - e^{-t}$ for $\beta = 2, 5, 10$. Compared to Table 6.12 when $\beta = n$, all T_1^* are less than those in Table 6.12, and T_2^* are less than those for small c_S/c_E and greater than those for large c_S/c_E. However, when c_S/c_E is large, both T_1^* and T_2^* in Table 6.13 are almost the same in Table 6.12, respectively, and also, when β is large, both T_1^* and T_2^* approach to those in Table 6.12. $\quad\square$

In addition, when T is a random variable with a general distribution $G(t)$ with mean $1/\theta$, from (6.78) and (6.84), the expected replacement cost is

$$C_1(G; \beta) = c_S \int_0^\infty [1 - e^{-\beta \overline{F}(t)}] G(t)\, dt + c_E \int_0^\infty e^{-\beta \overline{F}(t)} \overline{G}(t)\, dt, \tag{6.88}$$

and from (6.79) and (6.85), the expected cost rate is

$$C_2(G; \beta) = \frac{c_S \int_0^\infty [1 - e^{-\beta \overline{F}(t)}] G(t)\, dt + c_E \int_0^\infty e^{-\beta \overline{F}(t)} \overline{G}(t)\, dt}{\int_0^\infty [1 - e^{-\beta \overline{F}(t)}] \overline{G}(t)\, dt}. \tag{6.89}$$

In particular, when $G(t) = 1 - e^{-\theta t}$, (6.88) is

$$C_1(\theta; \beta) = c_S \int_0^\infty [1 - e^{-\beta \overline{F}(t)}]\, dt + \frac{c_E}{\theta} - (c_S + c_E) \int_0^\infty [1 - e^{-\beta \overline{F}(t)}] e^{-\theta t}\, dt, \tag{6.90}$$

and (6.89) is

$$C_2(\theta; \beta) = \frac{c_S \int_0^\infty [1 - e^{-\beta \overline{F}(t)}]\, dt + c_E/\theta}{\int_0^\infty [1 - e^{-\beta \overline{F}(t)}] e^{-\theta t}\, dt} - (c_S + c_E). \tag{6.91}$$

Optimum θ_i^* $(i = 1, 2)$ which minimize $C_i(\theta; \beta)$ satisfy the following respective equations:

$$\int_0^\infty [1 - e^{-\beta \overline{F}(t)}] \theta^2 t e^{-\theta t}\, dt = \frac{c_E}{c_S + c_E}, \tag{6.92}$$

$$\frac{\int_0^\infty [1 - e^{-\beta \overline{F}(t)}] e^{-\theta t}\, dt}{\int_0^\infty [1 - e^{-\beta \overline{F}(t)}] \theta^2 t e^{-\theta t}\, dt} - \frac{1}{\theta} = \frac{c_S}{c_E} \int_0^\infty [1 - e^{-\beta \overline{F}(t)}]\, dt. \tag{6.93}$$

Example 6.13 (Random replacement time) Table 6.14 presents optimum $1/\theta_i^*$ $(i = 1, 2)$ when $F(t) = 1 - e^{-t}$ for $\beta = 2, 5, 10$. Both optimum $1/\theta_i^*$ increase with c_S/c_E and β. This shows a similar tendency to Table 6.13. Compared to two tables, $1/\theta_i^*$ are less than T_i^* for small c_S/c_E and greater than T_i^* for large c_S/c_E. □

6.4.4 Nth Random Replacement

Suppose that the system operates for a job with random working times Y_j $(j = 1, 2, \ldots)$ discussed in Sect. 2.4. It is assumed that random variables Y_j are independent and have an identical distribution $G(t) \equiv \Pr\{Y_j \leq t\}$ with finite mean $1/\theta$ $(0 < \theta < \infty)$ in Fig. 2.2. In other words, the system operates for a job with a renewal process with an interarrival distribution $G(t)$. Let $G^{(j)}(t)$ $(j = 1, 2, \ldots)$ be the j-fold Stieltjes convolution of $G(t)$ with itself and $G^{(0)}(t) \equiv 1$ for $t \geq 0$. Suppose that the system is replaced before failure at the Nth $(N = 1, 2, \ldots)$ working time. Then, replacing $G(t)$ in (6.78) and (6.79) with $G^{(N)}(t)$, respectively, the

Table 6.14 Optimum $1/\theta_i^*$ ($i=1,2$) when $F(t)=1-e^{-t}$	c_S/c_E	$\beta=2$		$\beta=5$		$\beta=10$	
		$\frac{1}{\theta_1^*}$	$\frac{1}{\theta_2^*}$	$\frac{1}{\theta_1^*}$	$\frac{1}{\theta_2^*}$	$\frac{1}{\theta_1^*}$	$\frac{1}{\theta_2^*}$
	0.1	0.000	0.480	0.308	0.767	0.541	1.031
	0.2	0.054	0.655	0.455	1.012	0.721	1.335
	0.5	0.301	0.994	0.769	1.489	1.111	1.925
	1.0	0.587	1.372	1.164	2.020	1.603	2.583
	2.0	1.024	1.902	1.777	2.766	2.369	3.510
	5.0	1.964	2.950	3.103	4.243	4.020	5.345
	10.0	3.080	4.129	4.676	5.905	5.979	7.411
	20.0	4.697	6.795	6.958	8.253	8.817	10.331
	50.0	7.955	9.098	11.554	12.911	14.532	16.123

expected replacement cost is

$$C_1(N;n) = c_S \int_0^\infty [1 - F(t)^n] G^{(N)}(t) \, dt$$

$$+ c_E \int_0^\infty F(t)^n [1 - G^{(N)}(t)] \, dt \quad (N = 1, 2, \ldots), \tag{6.94}$$

and the expected cost rate is

$$C_2(N;n) = \frac{c_S \int_0^\infty [1 - F(t)^n] \, G^{(N)}(t) \, dt + c_E \int_0^\infty F(t)^n [1 - G^{(N)}(t)] \, dt}{\int_0^\infty [1 - F(t)^n][1 - G^{(N)}(t)] \, dt}$$

$$- (c_S + c_E) \quad (N = 1, 2, \ldots). \tag{6.95}$$

When $G(t) = 1 - e^{-\theta t}$, i.e., $G^{(N)}(t) = \sum_{j=N}^\infty [(\theta t)^j / j!] e^{-\theta t}$ ($N = 0, 1, 2, \ldots$), we find optimum N_i^* ($i = 1, 2$) which minimize $C_i(N;n)$ for a fixed $n \geq 1$, respectively. From the inequality $C_1(N+1;n) - C_1(N;n) \geq 0$,

$$\int_0^\infty F(t)^n \frac{\theta(\theta t)^N}{N!} e^{-\theta t} \, dt \geq \frac{c_S}{c_S + c_E},$$

or

$$\sum_{j=0}^N \int_0^\infty \frac{(\theta t)^j}{j!} e^{-\theta t} \, dF(t)^n \geq \frac{c_S}{c_S + c_E}, \tag{6.96}$$

Table 6.15 Optimum N_i^* ($i = 1, 2$) when $F(t) = 1 - e^{-t}$ and $G(t) = 1 - e^{-t}$

c_S/c_E	$n = 2$		$n = 5$		$n = 10$	
	N_1^*	N_2^*	N_1^*	N_2^*	N_1^*	N_2^*
0.1	1	1	1	1	1	1
0.2	1	1	1	1	1	1
0.5	1	1	1	1	2	2
1.0	1	1	2	2	3	3
2.0	2	2	3	3	3	4
5.0	3	3	4	4	5	5
10.0	4	4	5	5	6	6
20.0	5	5	6	6	7	7
50.0	6	6	7	7	8	8

whose left-hand side increases strictly with N to 1. Therefore, there exists a finite and unique minimum N_1^* ($1 \leq N_1^* < \infty$) which satisfies (6.96). Next, from the inequality $C_2(N + 1; n) - C_2(N; n) \geq 0$,

$$\frac{\sum_{j=0}^{N-1} \int_0^\infty [1 - F(t)^n][(\theta t)^j / j!] e^{-\theta t}\, dt}{\int_0^\infty [1 - F(t)^n][(\theta t)^N / N!] e^{-\theta t}\, dt} - N \geq \frac{c_S \theta \mu_n}{c_E},$$

or

$$\frac{\sum_{j=0}^{N-1} \int_0^\infty t[(\theta t)^j / j!] e^{-\theta t}\, dF(t)^n}{\sum_{j=N+1}^{\infty} \int_0^\infty [(\theta t)^j / j!] e^{-\theta t}\, dF(t)^n} \geq \frac{c_S \mu_n}{c_E}, \tag{6.97}$$

whose left-hand side increases strictly with N to ∞. Therefore, there exists a finite and unique minimum N_2^* ($1 \leq N_2^* < \infty$) which satisfies (6.97) (Problem 13 in Sect. 6.5).

Example 6.14 (Replacement for exponential working time) Suppose that $F(t) = 1 - e^{-t}$ and $G(t) = 1 - e^{-t}$. Then, from (6.96), N_1^* satisfies

$$\int_0^\infty (1 - e^{-t})^n \frac{t^N}{N!} e^{-t}\, dt \geq \frac{c_S}{c_S + c_E},$$

and from (6.97), N_2^* satisfies

$$\frac{\sum_{j=0}^{N-1} \int_0^\infty [1 - (1 - e^{-t})^n](t^j / j!) e^{-t}\, dt}{\int_0^\infty [1 - (1 - e^{-t})^n](t^N / N!) e^{-t}\, dt} - N \geq \frac{c_S}{c_E} \sum_{j=1}^{n} \frac{1}{j}.$$

Table 6.15 presents optimum N_i^* $(i = 1, 2)$ when $F(t) = 1 - e^{-t}$ and $G(t) = 1 - e^{-t}$. Both N_i^* $(i = 1, 2)$ increase with n and c_S/c_E and $N_1^* \le N_2^*$, however, N_1^* and N_2^* are almost the same. Compared to Table 6.12, N_i^* are also almost the same as T_i^* for small c_S/c_E. $\qquad\qquad\qquad\qquad\qquad\qquad\qquad\qquad\qquad\qquad\qquad\square$

6.5 Problems

1 Derive (6.2).
2 Derive (6.10) and (6.11) [16, p. 107].
3 Prove that the left-hand side of (6.26) increases strictly with β from 0 to ∞.
4 When N is a truncated Poisson distribution

$$\Pr\{N = n\} = \frac{1}{1 - e^{-\beta}} \frac{\beta^n}{n!} e^{-\beta} \quad (n = 1, 2, \ldots),$$

   compute β^* and $\tilde{\beta}$, and compare to Table 6.6.
5 Prove that $H_n(t)$ increases strictly with t from 0 to $h(\infty)$ for $n \ge 2$.
6 Prove that if $K < n$ and $(K/n)\sum_{j=K+1}^{n}(1/j) > c_1/c_F$ then there exists a finite and unique T_1^* which satisfies (6.56).
7 Prove that $\mu_{n,p}$ decreases with β from μ_n to $1/(n\lambda)$.
8 Compute $\mu_{n,p}$ and $\tilde{\mu}_{n,p}$ in Table 6.9.
9 Compute n^*, n_p^* and $\tilde{n}_p$ in Table 6.10.
10 Prove that $T_2^* \ge T_1^*$.
11 Prove that there exists a finite and unique θ_2^* which satisfies (6.83).
12 Prove that T_2^* increases strictly with β and $T_2^* \ge T_1^*$.
13 Consider the system with random number of units.

References

1. Barlow RE, Proschan F (1965) Mathematical theory of reliability. Wiley, New York
2. Ushakov IA (1994) Handbook of reliability engineering. Wiley, New York
3. Pham H (2003) Reliability of systems with multiple failure mode. In: Pham H (ed) Handbook of reliability engineering. Springer, London, pp 19–36
4. Blokus A (2006) Reliability analysis of large systems with dependent component. Inter J Reliab Qual Saf Eng 13:1–14
5. Zia L, Coit DW (2010) Redundancy allocation for series parallel systems using a column generation approach. IEEE Trans Reliab 59:706–717
6. Zuo MJ, Huang J, Kuo W (2003) Multi-state k-out-of-n systems. In: Pham H (ed) Handbook of reliability engineering. Springer, London, pp 3–17
7. Chang GJ, Cui L, Hwang FK (2000) Reliability of consecuitive-k systems. Kluwer, Dordrecht
8. Nakagawa T (2008) Advanced reliability models and maintenance policies. Springer, London
9. Nakagawa T, Yun WY (2011) Note on MTTF of a parallel system. Inter J Reliab Qual Saf Eng 18:1–8

10. Nakagawa T, Zhao X (2012) Optimization problems of a parallel system with a random number of units. IEEE Trans Reliab 61:543–548
11. Chen M, Mizutani S, Nakagawa T (2010) Random and age replacement policies. Inter J Reliab Qual Saf Eng 17:27–39
12. Nakagawa T, Zhao X, Yun WY (2011) Optimal age replacement and inspection policies with random failure and replacement time. Inter J reliab Qual Saf Eng 18:405–416
13. Zhao X, Nakagawa T (2012) Optimization problems of replacement first or last in reliability theory. Euro J Oper Res 223:141–149
14. Havil J (2003) GAMMA: Exploring Euler's constant. Princeton Univ Press, Princeton
15. Nakagawa T (2011) Stochastic processes with applications to reliability theory. Springer, London
16. Nakagawa T (2005) Maintenance theory of reliability. Springer, London
17. Kotz S, Nadarajah S (2000) Extreme value distribution. Imperial College, London

Chapter 7
Random Scheduling

Manufacturing systems in actual fields are subjected to many resources of uncertainty or randomness. Such uncertainty might result from machine failures and variable working times required for each job. Despite the fact that it is difficult to predict exactly a job completion time, the scheduling problem is still a major task in production management. A general overview of stochastic scheduling problems can be found [1–4].

The problem of scheduling jobs with random working times on a single machine has received significant attention due to its importance in developing scheduling theory and its practical aspect in regarding integrated manufacturing systems. Most literatures on a single machine problem have dealt with the determination of job sequences, and object criterions have been to optimize cost functions of performance measures: Optimum sequences that minimize the expected earliness and tardiness costs were discussed [5–7]. In contrast to optimum sequencing problems with random working times, scheduling models with consideration of machine failures have received little attention: The scheduling problem where jobs are subjected to shocks and can be successfully completed if no shock occurs during its working time was considered [8], and the problem with n jobs on a single machine subjected to failures was examined [9].

This chapter surveys some scheduling models [10, p. 82], [11, 12] and add new results by using reliability theory: Suppose that a job has a working time Y such as operating and processing times, and should be achieved in a scheduling time L. A job with random working times would be reasonable to assume that the completion time of a job is also a random variable Y. Section 7.1 derives a scheduling time in which a job is accomplished with some probability $1 - \varepsilon$. Furthermore, introducing the excess and shortage costs, we derive analytically an optimum single scheduling time that minimizes the total expected cost. In addition, we extend a single scheduling time to multiple scheduling times, using the inspection policy [13, p. 107], [14, p. 201].

Next, we consider the scheduling time L for two kinds of N random works: A tandem work is executed successively and a parallel work is executed at the same time. For such N works, we derive optimum scheduling times L^*, and conversely,

© Springer-Verlag London 2014
T. Nakagawa, *Random Maintenance Policies*,
Springer Series in Reliability Engineering, DOI 10.1007/978-1-4471-6575-0_7

optimum number N^* of units for given L. This would link the reliability scheduling by using two kinds of works to construct the framework of stochastic modeling.

System reliability can be improved by either redundancy or paralleling of units [10, p. 8], [15]. However, such redundant systems have not been used widely in the stochastic scheduling models. Another scheduling problem is how many number of units and what kinds of redundant systems we have to provide for N works. This is well known originally as the spare part problem [13, p. 49], in which how many number of spare parts should be provided to assure with some probability that a system will remain operating in time t. Section 7.2 takes up N tandem works operated on standby and parallel systems with n units and determines an optimum n^* for a job with N tandem works. Furthermore, bringing in shortage and excess costs, we derive an optimum number N^* for given n units, and conversely, derive optimum n^* units for given N works.

It has been assumed until now that the number N of works is constant. Suppose that the system operates for a job with works that arrive at a counter per day, per week, per month, and so on, which appears in the repairman problems [13, p. 139]. In this case, the number of works is not constant, however, it may be estimated statistically. Then, all results discussed in the previous sections are rewritten when N is a random variable with a discrete probability function. In particular, when a probability function is a Poisson distribution and a geometric distribution, optimum scheduling times L^* and optimum numbers n^* of units for standby and parallel systems are computed numerically in Sect. 7.3. Finally, as other reliability models, we will derive an optimum number N^* of works when the scheduling time is random in Sect. 8.4.1.

7.1 Scheduling of Random Works

We take up three scheduling times of a single work, N tandem works and N parallel works, and discuss their optimization problems.

7.1.1 Single Work

Suppose that a job has a working time Y. It would be better to assume that Y is a random variable with a general distribution $G(t) \equiv \Pr\{Y \le t\}$ with finite mean $1/\theta$ $(0 < \theta < \infty)$ and a density function $g(t)$, i.e., $1/\theta \equiv \int_0^\infty \overline{G}(t)\, dt$ and $g(t) \equiv dG(t)/dt$, where $\overline{\Phi}(t) \equiv 1 - \Phi(t)$ for any function $\Phi(t)$.

A job needs to be set up based on the scheduling time: If the work is not accomplished up to the scheduling time, its completion time is prolonged, and this causes much loss to scheduling. Conversely, if the work is completed too early before the scheduling time, this involves a waste of time or cost. The problem is how to determine an optimum scheduling time for a job with a random working time Y [10, p. 83]. It is assumed that the scheduling time for the work of a job is

Table 7.1 Optimum γL^* for $1 - \varepsilon$ when $G(t) = 1 - \exp[-(\gamma t)^\alpha]$

$1 - \varepsilon$	$\alpha = 1$	$\alpha = 2$	$\alpha = 3$
0.99	4.605	2.146	1.664
0.95	2.996	1.731	1.442
0.90	2.303	1.517	1.321
0.865	2.000	1.415	1.260
0.85	1.897	1.377	1.238
0.80	1.609	1.269	1.172
0.75	1.386	1.177	1.115
0.70	1.204	1.097	1.064

L $(0 \le L < \infty)$. First, for a given ε $(0 < \varepsilon < 1)$, we require a minimum L^* which satisfies

$$\Pr\{Y \le L\} = G(L) \ge 1 - \varepsilon. \tag{7.1}$$

Example 7.1 (Scheduling time for Weibull working time) When Y has a Weibull distribution $G(t) = 1 - \exp[-(\gamma t)^\alpha]$ $(\alpha \ge 1)$, from (7.1),

$$\gamma L^* = \left(\log \frac{1}{\varepsilon}\right)^{1/\alpha}. \tag{7.2}$$

Table 7.1 presents optimum γL^* given in (7.2) for different $1 - \varepsilon$. This indicates that γL^* increases with $1 - \varepsilon$, and decreases with α, because the randomness of working times may be lost with the increase in α. For example, when $\alpha = 1$ and $1 - \varepsilon = 0.99$, $\gamma L^* = 4.605$, i.e., L^* is 4.605 times the mean working time $1/\gamma$. When $\alpha = 1$ and $1 - \varepsilon = 0.865$, we should set up the scheduling time with two times of $1/\gamma$. □

Next, introduce the following costs: When the scheduling time is L, its cost is $c_0(L)$. If the work is accomplished up to time L, i.e., $L \ge Y$, it requires the excess cost $c_E(L - Y)$, and if it is not accomplished before time L and is completed after L, i.e., $L < Y$, it requires the shortage cost $c_S(Y - L)$ in Fig. 7.1. Then, the total expected cost until the work completion is

$$C(L) = \int_L^\infty c_S(t - L)\, dG(t) + \int_0^L c_E(L - t)\, dG(t) + c_0(L). \tag{7.3}$$

When $c_i(t) \equiv c_i t$ and $c_i > 0$ $(i = 0, S, E)$, the expected cost is

$$C(L) = c_S \int_L^\infty \overline{G}(t)\, dt + c_E \int_0^L G(t)\, dt + c_0 L. \tag{7.4}$$

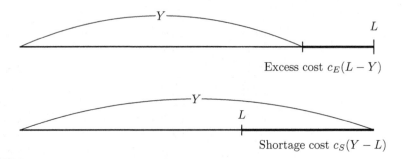

Fig. 7.1 Excess and shortage costs of scheduling a single work

Clearly,

$$C(0) \equiv \lim_{L \to 0} C(L) = \frac{c_S}{\theta}, \quad C(\infty) \equiv \lim_{L \to \infty} C(L) = \infty.$$

Thus, there exists a finite L^* ($0 \leq L^* < \infty$) which minimizes $C(L)$. Differentiating $C(L)$ with respect to L and setting it equal to zero,

$$G(L) = \frac{c_S - c_0}{c_S + c_E}. \tag{7.5}$$

Therefore, this corresponds to the same problem as (7.1). In particular, if $c_S \leq c_0$, then $L^* = 0$, i.e., we should not set up the scheduling of a job because its cost is too high. If L is not constant and is distributed exponentially with mean l, then (7.5) is written as

$$\int_0^\infty G(t) \frac{1}{l} e^{-t/l} \, dt = G^*(1/l) = \frac{c_S - c_0}{c_S + c_E}, \tag{7.6}$$

where $G^*(s)$ is the LS transform of $G(t)$ for $\mathrm{Re}(s) > 0$. Hence, when $G(t) = 1 - e^{-\theta t}$, l^* is given by

$$\theta l^* = \frac{c_S - c_0}{c_E + c_0}. \tag{7.7}$$

Next, consider the multiple scheduling times L_j ($j = 1, 2, \ldots$) in Fig. 7.2, where $T_j \equiv \sum_{i=1}^{j} L_i$ and $T_0 \equiv 0$: If we cannot accomplish the work at time T_{j-1}, we set up the next scheduling time L_j and continue the process until the work completion. Introduce the following costs: $c_1 + c_0 L_j$ is the jth scheduling cost, and $c_E(T_j - t)$ is the excess cost when the work is finished at time t. Then, the expect cost until the work completion is

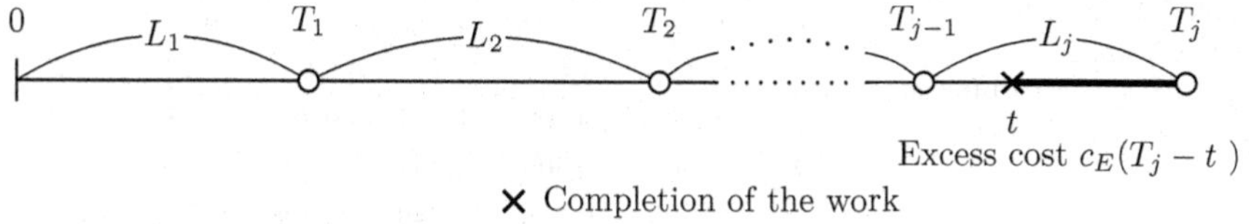

Fig. 7.2 Excess cost of multiple scheduling times

$$C(L_1, L_2, \ldots) = \sum_{j=1}^{\infty} \int_{T_{j-1}}^{T_j} [c_1 j + c_0 T_j + c_E(T_j - t)] \, dG(t)$$

$$= \sum_{j=0}^{\infty} [c_1 + (c_E + c_0)(T_{j+1} - T_j)] \overline{G}(T_j) - \frac{c_E}{\theta}. \qquad (7.8)$$

Differentiating $C(L_1, L_2, \ldots)$ with respect to T_j and setting it equal to zero (Problem 1 in Sect. 7.4),

$$T_{j+1} - T_j = \frac{G(T_j) - G(T_{j-1})}{g(T_j)} - \frac{c_1}{c_E + c_0} \quad (j = 1, 2, \ldots). \qquad (7.9)$$

Recall that (7.9) corresponds to the type of the equation [13, p. 110], [14, p. 203] for the inspection policy in (4.17) of Chap. 4. Thus, we can compute optimum scheduled times L_j and T_j, using Algorithm of [13, p. 112].

In particular, when $G(t) = 1 - e^{-\theta t}$, (7.9) is

$$L_{j+1} = \frac{e^{\theta L_j} - 1}{\theta} - \frac{c_1}{c_E + c_0}. \qquad (7.10)$$

Let L_1 be a solution of the equation

$$L_1 = \frac{e^{\theta L_1} - 1}{\theta} - \frac{c_1}{c_E + c_0},$$

i.e.,

$$\frac{e^{\theta L_1} - 1}{\theta} - L_1 = \frac{c_1}{c_E + c_0}, \qquad (7.11)$$

whose left-hand side increases strictly with L_1 from 0 to ∞. Thus, there exists a finite and unique $L^* \equiv L_1$ ($0 < L^* < \infty$) which satisfies (7.11). Therefore, we can easily obtain $L_j \equiv L^*$ for all $j \geq 1$ from (7.10) (Problem 2 in Sect. 7.4). Clearly, L^* increases strictly with $1/\theta$ from 0 to ∞. In this case, we set up a constant scheduling time L^* for any scheduling. This can be obviously seen because an

Table 7.2 Optimum γT_j^*
and γL_j^* when
$\gamma c_1/(c_E + c_0) = 0.1487$ and
$G(t) = 1 - \exp[-(\gamma t)^\alpha]$

j	$\alpha = 1$		$\alpha = 2$		$\alpha = 3$	
	γT_j^*	γL_j^*	γT_j^*	γL_j^*	γT_j^*	γL_j^*
1	0.500	0.500	0.767	0.767	0.893	0.893
2	1.000	0.500	1.141	0.374	1.177	0.284
3	1.500	0.500	1.449	0.308	1.392	0.215
4	2.000	0.500	1.720	0.271	1.572	0.180
5	2.500	0.500	1.967	0.247	1.728	0.156
6	3.000	0.500	2.197	0.230	1.868	0.140
7	3.500	0.500	2.413	0.216	1.996	0.128
8	4.000	0.500	2.618	0.205	2.114	0.118
9	4.500	0.500	2.815	0.197	2.225	0.111
10	5.000	0.500	3.006	0.191	2.330	0.105

exponential distribution has a memoryless property in stochastic processes [16, p. 23], [17, p. 13].

Example 7.2 (Sequential scheduling time) Table 7.2 presents optimum T_j^* and L_j^* for $\alpha = 1, 2, 3$ when $\gamma c_1/(c_E + c_0) = 0.1478$ and $G(t) = 1 - \exp[-(\gamma t)^\alpha]$. Note that when $\alpha = 1$, $\gamma = \theta$ and $\gamma L_j^* = 0.5$ is constant and is calculated from (7.11). When $\alpha > 1$, γL_j^* decrease with j and are <0.5 for any $j \geq 2$. It is of interest that γT_j^* increases with α for $j \leq 2$ and decreases with α for $j \geq 3$ (Problem 3 in Sect. 7.4).   □

7.1.2 N Tandem Works

Consider the scheduling problems of N works ($N = 1, 2, \ldots$) in tandem in Fig. 7.3: It is assumed that Y_j ($j = 1, 2, \ldots, N$) are the working times for a job, and are independent and have an identical distribution $\Pr\{Y_j \leq t\} \equiv G(t)$ with finite mean $1/\theta$, which is called N tandem works, where $S_N \equiv \sum_{j=1}^{N} Y_j$. Then, the probability that N works finish until time L is

$$\Pr\{Y \leq L\} = \Pr\{Y_1 + Y_2 + \cdots + Y_N \leq L\} = G^{(N)}(L),$$

where $\Phi^{(j)}(t)$ ($j = 1, 2, \ldots$) denotes the j-fold Stieltjes convolution of any function $\Phi(t)$ and $\Phi^{(0)}(t) \equiv 1$ for $t \geq 0$.

The following items of costs are introduced the same as those in Sect. 7.1.1: The scheduling cost is $c_0 L$, the shortage cost is $c_S (S_N - L)$ and the excess cost is $c_E (L - S_N)$. By replacing $G(t)$ in (7.4) with $G^{(N)}(t)$, the total expected cost until the N work completion is

Fig. 7.3 Process of N tandem works

$$C_T(L, N) = c_S \int_L^\infty [1 - G^{(N)}(t)] \, dt + c_E \int_0^L G^{(N)}(t) \, dt + c_0 L. \qquad (7.12)$$

Thus, an optimum interval L_{T1}^* for given $N \geq 1$ is, from (7.5),

$$G^{(N)}(L) = \frac{c_S - c_0}{c_S + c_E}. \qquad (7.13)$$

It can be seen that L_{T1}^* increases strictly with N.

Next, we find an optimum L_{T2}^* for given $N \geq 1$ which minimizes the following expected cost rate:

$$C_T(L) \equiv \frac{C_T(L, N)}{L} = \frac{c_S \int_L^\infty [1 - G^{(N)}(t)] \, dt + c_E \int_0^L G^{(N)}(t) \, dt}{L} + c_0. \qquad (7.14)$$

Clearly,

$$C_T(0) \equiv \lim_{L \to 0} C_T(L) = \infty, \quad C_T(\infty) \equiv \lim_{L \to \infty} C_T(L) = c_0 + c_E.$$

Differentiating $C_T(L)$ with respect to L and setting it equal to zero,

$$\frac{\theta}{N} \int_0^L [G^{(N)}(L) - G^{(N)}(t)] \, dt = \frac{c_S}{c_S + c_E}, \qquad (7.15)$$

whose left-hand increases strictly with L from 0 to 1. Therefore, there exists a finite and unique L_{T2}^* $(0 < L_{T2}^* < \infty)$ which satisfies (7.15).

Conversely, we obtain an optimum number N_T^* of works which minimizes $C_T(L, N)$ in (7.12) for given $L > 0$. From the inequality $C_T(L, N + 1) - C_T(L, N) \geq 0$,

$$\frac{\int_L^\infty [G^{(N)}(t) - G^{(N+1)}(t)] \, dt}{\int_0^L [G^{(N)}(t) - G^{(N+1)}(t)] \, dt} \geq \frac{c_E}{c_S},$$

i.e.,

$$\theta \int_0^L [G^{(N)}(t) - G^{(N+1)}(t)] \, dt \leq \frac{c_S}{c_S + c_E} \quad (N = 0, 1, 2, \ldots). \tag{7.16}$$

In particular, when $G(t) = 1 - e^{-\theta t}$,

$$\sum_{j=0}^{N} \frac{(\theta L)^j}{j!} e^{-\theta L} \geq \frac{c_E}{c_S + c_E}. \tag{7.17}$$

Thus, an optimum N_T^* $(0 \leq N_T^* < \infty)$ is given by a finite and unique minimum which satisfies (7.17) and increases with L from 0 to ∞. If $1 - e^{-\theta L} \leq c_S/(c_S + c_E)$, then $N_T^* = 0$, i.e., we should not set up any work because L is too short to do it. Furthermore, we could compute both L_{T2}^* and N_T^* which minimize the expected cost rate (Problem 4 in Sect. 7.4)

$$\widetilde{C}_T(L, N) = \frac{1}{\theta L} \left[c_S \sum_{j=0}^{N} (N - j) \frac{(\theta L)^j}{j!} e^{-\theta L} + c_E \sum_{j=N}^{\infty} (j - N) \frac{(\theta L)^j}{j!} e^{-\theta L} \right] + c_0.$$

7.1.3 N Parallel Works

Suppose that N works begin to operate at the same time from $t = 0$, which is called N parallel works. Then, by replacing $G^{(N)}(t)$ in (7.12) with $G(t)^N$, the total expected cost is

$$C_P(L, N) = c_S \int_L^{\infty} [1 - G(t)^N] \, dt + c_E \int_0^L G(t)^N \, dt + c_0 L. \tag{7.18}$$

The optimum L_{P1}^* for a given $N \geq 1$ is, from (7.13),

$$G(L) = \left(\frac{c_S - c_0}{c_S + c_E} \right)^{1/N}. \tag{7.19}$$

It can be seen that L_{P1}^* increases strictly with N and is less than L_{T1}^* for N tandem works, because $G(L)^N \geq G^{(N)}(L)$.

Next, we find an optimum L_{P2}^* for given $N \geq 1$ which minimizes the following expected cost rate:

$$C_P(L) \equiv \frac{C_P(L, N)}{L} = \frac{c_S \int_L^\infty [1 - G(t)^N] \, dt + c_E \int_0^L G(t)^N \, dt}{L} + c_0. \quad (7.20)$$

Differentiating $C_P(T)$ with respect to L and setting it equal to zero,

$$\frac{\int_0^L [G(L)^N - G(t)^N] \, dt}{\int_0^\infty [1 - G(t)^N] \, dt} = \frac{c_S}{c_S + c_E}, \quad (7.21)$$

whose left-hand increases strictly from 0 to 1. Therefore, there exists a finite and unique L_{P2}^* $(0 < L_{P2}^* < \infty)$ which satisfies (7.21). In particular, when $G(t) = 1 - e^{-\theta t}$, (7.21) is (Problem 5 in Sect. 7.4)

$$\frac{\sum_{j=1}^N [(1 - e^{-\theta L})^j / j] - \theta L[1 - (1 - e^{-\theta L})^N]}{\sum_{j=1}^N (1/j)} = \frac{c_S}{c_S + c_E}. \quad (7.22)$$

Conversely, an optimum N_P^* for given L $(0 < L < \infty)$ is, from (7.16),

$$\frac{\int_L^\infty G(t)^N \overline{G}(t) \, dt}{\int_0^L G(t)^N \overline{G}(t) \, dt} \geq \frac{c_E}{c_S}. \quad (7.23)$$

Because

$$\int_L^\infty G(t)^{N+1} \overline{G}(t) \, dt \int_0^L G(t)^N \overline{G}(t) \, dt - \int_L^\infty G(t)^N \overline{G}(t) \, dt \int_0^L G(t)^{N+1} \overline{G}(t) \, dt$$

$$\geq G(L) \left[\int_L^\infty G(t)^N \overline{G}(t) \, dt \int_0^L G(t)^N \overline{G}(t) \, dt - \int_L^\infty G(t)^N \overline{G}(t) \, dt \int_0^L G(t)^N \overline{G}(t) \, dt \right] = 0,$$

the left-hand side of (7.23) increases strictly with N. In addition,

$$\lim_{N \to \infty} \frac{\int_L^\infty G(t)^N \overline{G}(t) \, dt}{\int_0^L G(t)^N \overline{G}(t) \, dt} \geq \lim_{N \to \infty} \frac{G(L)^N \int_L^\infty \overline{G}(t) \, dt}{\int_0^L G(t)^N \overline{G}(t) \, dt}$$

$$= \lim_{N \to \infty} \frac{\int_L^\infty \overline{G}(t) \, dt}{\int_0^L [G(t)/G(L)]^N \overline{G}(t) \, dt} = \infty.$$

The left-hand side of (7.23) decreases with L. Thus, the left-hand side of (7.23) increases strictly with N to ∞ and decreases with L. Therefore, there exists a finite and unique minimum N_P^* $(0 \leq N_P^* < \infty)$ which satisfies (7.23) and increases with L from 0 to ∞.

When $G(t) = 1 - e^{-\theta t}$, an optimum N_P^* $(0 \leq N_P^* < \infty)$ is given by a finite and unique minimum which satisfies

$$(1 - e^{-\theta L})^{N+1} \leq \frac{c_S}{c_S + c_E}. \tag{7.24}$$

If $1 - e^{-\theta L} \leq c_S/(c_S + c_E)$, then $N_P^* = 0$, which is the same as Sect. 7.2.1. Furthermore, we could derive both L_{P2}^* and N_P^* which minimize the expected cost rate (Problem 4 in Sect. 7.4)

$$\tilde{C}_P(L, N) = \frac{1}{\theta L} \left\{ c_S \sum_{j=1}^{N} \left[1 - \frac{(1 - e^{-\theta L})^j}{j} \right] + c_E \sum_{j=N+1}^{\infty} \frac{(1 - e^{-\theta L})^j}{j} \right\} + c_0.$$

7.2 Redundant Systems

Suppose that redundant systems with n ($n = 1, 2, \ldots$) units operate for a job with N ($1 \leq N < \infty$) works. It is assumed that each unit is independent and has an identical failure distribution $F(t)$ with finite mean $1/\lambda$ ($0 < \lambda < \infty$). The redundant systems with n units need to take an operating cost $c_0 n$. In this section, we adopt a standard standby system and parallel system for a job.

7.2.1 Standby System

We consider the same cost structure introduced in Sect. 7.1: If a standby system with n units fails at time t and Nth work finishes at time $u(<t)$, it requires the excess cost $c_E(t - u)$, and conversely, if the system fails at time u and the work finishes at time $t(>u)$, it requires the shortage cost $c_S(t - u)$. Then, adding the operating cost $c_0 n$, the total expected cost is

$$C_S(n, N) = c_S \int_0^\infty \left[\int_0^t (t - u) \, dF^{(n)}(u) \right] dG^{(N)}(t)$$

$$+ c_E \int_0^\infty \left[\int_0^t (t - u) dG^{(N)}(u) \right] dF^{(n)}(t) + c_0 n$$

$$= c_S \int_0^\infty F^{(n)}(t)[1 - G^{(N)}(t)] dt$$

$$+ c_E \int_0^\infty [1 - F^{(n)}(t)]G^{(N)}(t) dt + c_0 n$$

$$= c_S \frac{N}{\theta} + c_E \frac{n}{\lambda} - (c_S + c_E)$$

$$\times \int_0^\infty [1 - F^{(n)}(t)][1 - G^{(N)}(t)]dt + c_0 n \quad (n, N = 1, 2, \ldots). \quad (7.25)$$

In particular, when $G(t) = 1 - e^{-\theta t}$ and $F(t) = 1 - e^{-\lambda t}$, the expected cost is (Problem 6 in Sect. 7.4)

$$C_S(n, N) = c_S \frac{N}{\theta} + c_E \frac{n}{\lambda}$$

$$- \frac{c_S + c_E}{\theta + \lambda} \sum_{j=0}^{n-1} \sum_{i=0}^{N-1} \binom{i+j}{i} \left(\frac{\lambda}{\theta + \lambda}\right)^j \left(\frac{\theta}{\theta + \lambda}\right)^i + c_0 n. \quad (7.26)$$

We find an optimum n_S^* for given $N \geq 1$ and N_S^* for given $n \geq 1$, respectively. From the inequality $C_S(n + 1, N) - C_S(n, N) \geq 0$,

$$\left(\frac{\lambda}{\theta + \lambda}\right)^{n+1} \sum_{j=0}^{N-1} \binom{n+j}{j} \left(\frac{\theta}{\theta + \lambda}\right)^j \leq \frac{c_E + \lambda c_0}{c_S + c_E} \quad (n = 1, 2, \ldots). \quad (7.27)$$

Letting $L_S(n)$ be the left-hand side of (7.27),

$$L_S(n) - L_S(n + 1) = \binom{n+N}{N-1} \left(\frac{\theta}{\theta + \lambda}\right)^N \left(\frac{\lambda}{\theta + \lambda}\right)^{n+1} > 0.$$

Thus, $L_S(n)$ decreases strictly with n to 0. Therefore, there exists a finite and unique minimum n_S^* ($1 \leq n_S^* < \infty$) which satisfies (7.27) and increases with N. In particular, when $N = 1$, an optimum n_S^* is given by an integer which satisfies

$$\log \left(\frac{c_E + c_0 \lambda}{c_S + c_E}\right) \Big/ \log \left(\frac{\lambda}{\theta + \lambda}\right) - 1 \leq n_S^* < \log \left(\frac{c_E + c_0 \lambda}{c_S + c_E}\right) \Big/ \log \left(\frac{\lambda}{\theta + \lambda}\right).$$

Next, from the inequality $C_S(n, N + 1) - C_S(n, N) \geq 0$,

$$\left(\frac{\theta}{\theta + \lambda}\right)^{N+1} \sum_{j=0}^{n-1} \binom{N+j}{j} \left(\frac{\lambda}{\theta + \lambda}\right)^j \leq \frac{c_S}{c_S + c_E} \quad (N = 1, 2, \ldots), \quad (7.28)$$

whose left-hand side decreases strictly with N to 0. Therefore, there exists a finite and unique minimum N_S^* ($1 \leq N_S^* < \infty$) which satisfies (7.28) and increases with n. In particular, when $n = 1$, optimum N_S^* is given by an integer which satisfies

Table 7.3 Optimum n_S^* for N and N_S^* for n when $c_E/c_S = 0.5$ and $\lambda c_0/c_S = 0.1$

λ/θ	N					n				
	1	2	3	4	5	1	2	3	4	5
0.1	1	1	1	1	1	4	11	20	29	37
0.2	1	1	1	1	1	2	6	10	14	18
0.3	1	1	1	1	2	1	4	6	9	12
0.4	1	1	1	2	2	1	3	5	7	9
0.5	1	1	2	2	3	1	2	4	5	7
0.6	1	1	2	3	3	1	2	3	4	6
0.7	1	1	2	3	4	1	1	2	4	5
0.8	1	2	2	3	4	1	1	2	3	4
0.9	1	2	3	4	5	1	1	2	3	4

$$\log\left(\frac{c_S}{c_S + c_E}\right) \Big/ \log\left(\frac{\theta}{\theta + \lambda}\right) - 1 \leq N_S^* < \log\left(\frac{c_S}{c_S + c_E}\right) \Big/ \log\left(\frac{\theta}{\theta + \lambda}\right).$$

Example 7.3 (*Number for exponential failure and working times*) Table 7.3 presents optimum n_S^* and N_S^* for different N, n and λ/θ, respectively, when $c_E/c_S = 0.5$ and $\lambda c_0/c_S = 0.1$. Optimum n_S^* increases with λ/θ and N, but N_S^* decreases with λ/θ and increases with n. It is of interest that when $\lambda/\theta = 0.9$, $n_S^* = N$, and n_S^* is almost equal to $[N\lambda/\theta + 1]$ and N_S^* is almost equal to $[(n-1)/(\lambda/\theta)]$ for $n \geq 2$, where $[x]$ denotes the greatest integer contained in x In other words, the mean time to the completion of N works is almost the same mean failure time for a standby system with $(n-1)$ units (Problem 7 in Sect. 7.4). □

7.2.2 Parallel System

We consider a parallel system with n units which has a linear cost structure. Then, replacing $F^{(n)}(t)$ in (7.25) with $F(t)^n$ formally, the total expected cost is

$$C_P(n, N) = c_S \int_0^\infty F(t)^n [1 - G^{(N)}(t)] \, dt$$

$$+ c_E \int_0^\infty [1 - F(t)^n] G^{(N)}(t) \, dt + c_0 n \quad (n, N = 1, 2, \ldots). \quad (7.29)$$

In particular, when $G(t) = 1 - e^{-\theta t}$ and $F(t) = 1 - e^{-\lambda t}$, the expected cost is (Problem 6 in Sect. 7.4)

$$C_P(n, N) = c_S \frac{N}{\theta} + \frac{c_E}{\lambda} \sum_{j=1}^{n} \frac{1}{j} + c_0 n$$

$$- (c_S + c_E) \sum_{j=0}^{N-1} \int_0^\infty [1 - (1 - e^{-\lambda t})^n] \frac{(\theta t)^j}{j!} e^{-\theta t} \, dt$$

$$= c_S \frac{N}{\theta} + \frac{c_E}{\lambda} \sum_{j=1}^{n} \frac{1}{j} + c_0 n$$

$$+ \frac{c_S + c_E}{\lambda} \sum_{j=1}^{n} (-1)^j \binom{n}{j} \frac{1}{j} \left[1 - \left(\frac{\theta}{j\lambda + \theta} \right)^N \right]. \qquad (7.30)$$

From the inequality $C_P(n + 1, N) - C_P(n, N) \geq 0$,

$$\sum_{j=0}^{n} (-1)^j \binom{n}{j} \frac{1}{j+1} \left\{ 1 - \left[\frac{\theta}{(j+1)\lambda + \theta} \right]^N \right\} \leq \frac{c_E/(n+1) + c_0\lambda}{c_S + c_E} \quad (n = 1, 2, \ldots).$$

$$(7.31)$$

Letting $L_P(n)$ be the left-hand side of (7.31),

$$L_P(n) - L_P(n+1) = \sum_{j=0}^{n} (-1)^j \binom{n}{j} \frac{1}{j+2} \left\{ 1 - \left[\frac{\theta}{(j+2)\lambda + \theta} \right]^N \right\} > 0.$$

Thus, $L_P(n)$ decreases strictly with n to 0, and the right-hand side of (7.31) decreases to $c_0\lambda/(c_S + c_E)$. Therefore, there exists a finite and unique minimum n_P^* ($1 \leq n_P^* < \infty$) which satisfies (7.31) and increases with N.

Next, from the inequality $C_P(n, N+1) - C_P(n, N) \geq 0$,

$$\sum_{j=0}^{n} (-1)^{j+1} \binom{n}{j} \left(\frac{\theta}{j\lambda + \theta} \right)^{N+1} \leq \frac{c_S}{c_S + c_E} \quad (N = 1, 2, \ldots), \qquad (7.32)$$

whose left-hand side decreases strictly with N to 0. Therefore, there exists a finite and unique minimum N_P^* ($1 \leq N_P^* < \infty$) which satisfies (7.32) and increases with n.

Example 7.4 (Number for exponential failure and working times) Table 7.4 presents optimum n_P^* and N_P^* for different N, n and λ/θ, respectively, when $c_E/c_S = 0.5$ and $\lambda c_0/c_S = 0.1$ and shows a similar tendency to Table 7.3. Notice that when $\lambda/\theta = 0.7$, $n_P^* = N$. Compared to Table 7.3, $n_P^* \geq n_S^*$ and $N_P^* \leq N_S^*$, because a parallel system needs more units than a standby system for the same job. It is of interest that when $\lambda/\theta = 0.9$, $n_S^* = N$, and n_S^* is almost equal to a minimum such that $\sum_{j=1}^{n} (1/j) \geq N(\lambda/\theta)$, and N_S^* is almost equal to a minimum such that

Table 7.4 Optimum n_p^* for N and N_p^* for n when $c_E/c_S = 0.5$ and $\lambda c_0/c_S = 0.1$

λ/θ	N					n				
	1	2	3	4	5	1	2	3	4	5
0.1	1	1	1	1	1	4	8	11	14	16
0.2	1	1	1	1	1	2	4	5	7	8
0.3	1	1	1	1	2	1	2	4	4	5
0.4	1	1	1	2	3	1	2	3	3	4
0.5	1	1	2	3	4	1	1	2	2	3
0.6	1	1	2	3	5	1	1	2	2	2
0.7	1	2	3	4	5	1	1	1	2	2
0.8	1	2	3	5	6	1	1	1	1	2
0.9	1	2	4	5	7	1	1	1	1	1

$N \geq \sum_{j=1}^{n}(1/j)/(\lambda/\theta)$. In this case, the mean time to the completion of N works is almost the same failure time for a parallel system with n units (Problem 8 in Sect. 7.4).

$\square$

Furthermore, we could discuss both optimum n^* and N^* which minimize $C_S(n, N)$ in (7.25) and $C_P(n, N)$ in (7.29), (Problem 9 in Sect. 7.4) respectively.

7.3 Random Number of Works

It has been assumed that the number N of works is constant and is previously given. Suppose that a system operates for a job with a variable number of works which arrive at a counter per day, per week, per month, and so on, as shown in queueing processes [18] and repairman problems [13, p. 139]. In this case, the number of works is not constant, however, it may be estimated statistically. It is assumed that the number N of works is a random variable with a discrete probability function $p_k \equiv \Pr\{N = k\}$ $(k = 1, 2, \ldots)$. Then, all results in the previous sections can be rewritten.

7.3.1 N Tandem Works

Consider the scheduling problems of N tandem works when N has a probability function p_k $(k = 1, 2, \ldots)$ with finite mean μ $(0 < \mu < \infty)$. Then, the total expected cost until the N work completion is, from (7.12),

$$C_T(L; p_k) = c_S \sum_{k=1}^{\infty} p_k \int_L^{\infty} [1 - G^{(k)}(t)] \, dt + c_E \sum_{k=1}^{\infty} p_k \int_0^L G^{(k)}(t) \, dt + c_0 L.$$

$$(7.33)$$

Thus, an optimum interval L_{T1}^* which minimizes (7.33) is, from (7.13),

$$\sum_{k=1}^{\infty} p_k G^{(k)}(L) = \frac{c_S - c_0}{c_S + c_E}.$$

$$(7.34)$$

Furthermore, the expected cost rate is, from (7.14),

$$\begin{aligned} C_T(L) &\equiv \frac{C_T(L; p_k)}{L} \\ &= \frac{c_S \sum_{k=1}^{\infty} p_k \int_L^{\infty} [1 - G^{(k)}(t)] \, dt + c_E \sum_{k=1}^{\infty} p_k \int_0^L G^{(k)}(t) \, dt}{L} + c_0. \end{aligned}$$

$$(7.35)$$

An optimum L_{T2}^* which minimizes (7.35) is, from (7.15),

$$\frac{\theta}{\mu} \sum_{k=1}^{\infty} p_k \int_0^L [G^{(k)}(L) - G^{(k)}(t)] \, dt = \frac{c_S}{c_S + c_E}.$$

$$(7.36)$$

Example 7.5 (Scheduling interval for Poisson and geometric distributions) When $p_k = [\beta^{k-1}/(k-1)!] e^{-\beta}$ $(k = 1, 2, \ldots)$, i.e., $\mu = \beta + 1$, and $G(t) = 1 - e^{-\theta t}$, (7.34) is

$$\sum_{k=0}^{\infty} \frac{\beta^k}{k!} e^{-\beta} \sum_{j=k+1}^{\infty} \frac{(\theta L)^j}{j!} e^{-\theta L} = \frac{c_S - c_0}{c_S + c_E},$$

$$(7.37)$$

and (7.36) is

$$\frac{1}{\beta + 1} \sum_{k=0}^{\infty} \frac{\beta^k}{k!} e^{-\beta} \left[\sum_{j=0}^{k} \sum_{i=j+1}^{\infty} \frac{(\theta L)^i}{i!} e^{-\theta L} - \sum_{j=0}^{k} \frac{(\theta L)^{j+1}}{j!} e^{-\theta L} \right] = \frac{c_S}{c_S + c_E}.$$

$$(7.38)$$

Next, when $p_k = pq^{k-1}$ $(k = 1, 2, \ldots)$ and $G(t) = 1 - e^{-\theta t}$, (7.34) is

$$1 - e^{-p\theta L} = \frac{c_S - c_0}{c_S + c_E},$$

$$(7.39)$$

Table 7.5 Optimum θL_{T1}^*, θL_{T2}^*, θL_{R1}^*, θL_{R2}^*, and θL_{G1}^*, θL_{G2}^* when $c_E/c_S = 0.5$ and $c_0/c_S = 0.1$

N	θL_{T1}^*	θL_{T2}^*	θL_{R1}^*	θL_{R2}^*	θL_{G1}^*	θL_{G2}^*
1	0.916	2.289	0.916	2.289	0.916	2.289
2	2.022	3.433	1.973	4.045	1.833	4.579
5	5.237	6.753	5.280	7.818	4.581	11.446
10	10.476	12.134	10.626	13.459	9.163	22.893
20	20.811	22.685	21.106	24.292	18.326	45.786
50	51.473	53.793	52.047	55.894	45.815	114.464
100	102.217	105.049	103.102	107.683	91.629	228.928

and when $\mu = 1/p$, (7.36) is

$$1 - (1 + p\theta L)e^{-p\theta L} = \frac{c_S}{c_S + c_E}. \tag{7.40}$$

Table 7.5 presents optimum L_{T1}^* and L_{T2}^* which satisfy (7.13) and (7.15), L_{R1}^* and L_{R2}^* which satisfy (7.37) and (7.38) when $\beta + 1 = N$, and L_{G1}^* and L_{G2}^* which satisfy (7.39) and (7.40) when $p = 1/N$, respectively. This indicates that $L_{i1}^* < L_{i2}^*$ ($i = T, R, G$) and $L_{T2}^* < L_{i2}^*$ ($i = R, G$) for $N \geq 2$. From (7.39) and (7.40), $\theta L_{G1}^*/N = 0.916$ and $\theta L_{G2}^*/N = 2.289$ for all N.   □

7.3.2 N Parallel Works

Consider the scheduling problems of N parallel works when N has a probability function p_k ($k = 1, 2, \ldots$). Then, by replacing $G^{(k)}(t)$ in (7.33) with $G(t)^k$, the total expected cost is

$$C_P(L; p_k) = c_S \sum_{k=1}^{\infty} p_k \int_L^{\infty} [1 - G(t)^k] \, dt + c_E \sum_{k=1}^{\infty} p_k \int_0^L G(t)^k \, dt + c_0 L. \tag{7.41}$$

Thus, an optimum L_{R1}^* which minimizes (7.41) is, from (7.34),

$$\sum_{k=1}^{\infty} p_k G(L)^k = \frac{c_S - c_0}{c_S + c_E}. \tag{7.42}$$

Furthermore, the expected cost rate is, from (7.35),

$$C_P(L) \equiv \frac{C_P(L; p_k)}{L}$$

$$= \frac{c_S \sum_{k=1}^{\infty} p_k \int_L^{\infty} [1 - G(t)^k] dt + c_E \sum_{k=1}^{\infty} p_k \int_0^L G(t)^k dt}{L} + c_0. \quad (7.43)$$

An optimum L_{R2}^* which minimizes (7.43) is, from (7.36),

$$\frac{\sum_{k=1}^{\infty} p_k \int_0^L [G(L)^k - G(t)^k] dt}{\sum_{k=1}^{\infty} p_k \int_0^{\infty} [1 - G(t)^k] dt} = \frac{c_S}{c_S + c_E}. \quad (7.44)$$

Example 7.6 (Scheduling for Poisson and geometric distributions) When $p_k = [\beta^{k-1}/(k-1)!] e^{-\beta}$ $(k = 1, 2, \ldots)$ and $G(t) = 1 - e^{-\theta t}$, (7.42) is

$$(1 - e^{-\theta L}) \exp(-\beta e^{-\theta L}) = \frac{c_S - c_0}{c_S + c_E}, \quad (7.45)$$

and (7.44) is

$$\frac{\int_0^L [(1 - e^{-\theta L}) \exp(-\beta e^{-\theta L}) - (1 - e^{-\theta t}) \exp(-\beta e^{-\theta t})] dt}{\int_0^{\infty} [1 - (1 - e^{-\theta t}) \exp(-\beta e^{-\theta t})] dt} = \frac{c_S}{c_S + c_E}. \quad (7.46)$$

Next, when $p_k = pq^{k-1}$ $(k = 1, 2, \ldots)$ and $G(t) = 1 - e^{-\theta t}$, (7.42) is

$$\frac{e^{-\theta L}}{p + q e^{-\theta L}} = \frac{c_E + c_0}{c_S + c_E}, \quad (7.47)$$

and (7.44) is

$$\frac{\int_0^L [1/(p + q e^{-\theta L}) - 1/(p + q e^{-\theta t})] dt}{\int_0^{\infty} [1/p - 1/(p + q e^{-\theta t})] dt} = \frac{c_S}{c_S + c_E}. \quad (7.48)$$

Table 7.6 presents optimum L_{P1}^* and L_{P2}^* which satisfy (7.19) and (7.22), L_{R1}^* and L_{R2}^* which satisfy (7.45) and (7.46) when $N = \beta + 1$, and L_{G1}^* and L_{G2}^* which satisfy (7.47) and (7.48) when $N = 1/p$, respectively. This indicates that $L_{i1}^* \leq L_{i2}^*$ $(i = P, R, G)$, $L_{P1}^* \geq L_{i1}^*$ $(i = R, G)$, however, $L_{P2}^* < L_{i2}^*$ $(i = R, G)$, and $L_{R1}^* \geq L_{G1}^*$, however, $L_{R2}^* \leq L_{G2}^*$. Furthermore, both L_{P1}^* and L_{R1}^*, and both L_{P2}^* and L_{R2}^* are almost the same for large N. Naturally, compared to Table 7.5, all values in Table 7.6 are less than those in Table 7.5. $\square$

Table 7.6 Optimum θL_{P1}^*, θL_{P2}^*, θL_{R1}^*, θL_{R2}^*, and θL_{G1}^*, θL_{G2}^* when $c_E/c_S = 0.5$ and $c_0/c_S = 0.1$

N	θL_{P1}^*	θL_{P2}^*	θL_{R1}^*	θL_{R2}^*	θL_{G1}^*	θL_{G2}^*
1	0.916	2.289	0.916	2.289	0.916	2.289
2	1.490	2.650	1.434	2.701	1.386	2.726
5	2.332	3.266	2.292	3.293	2.140	3.370
10	3.000	3.809	2.977	3.818	2.773	3.902
20	3.680	4.395	3.668	4.398	3.434	4.465
50	4.589	5.214	4.584	5.214	4.331	5.248
100	5.279	5.854	5.227	5.854	5.017	5.863

7.3.3 Standby System

Consider a standby system with n units for a random number N of works, where each unit has a failure distribution $F(t)$ and N is a random variable with a probability function $p_k \equiv \Pr\{N = k\}$ $(k = 1, 2, \ldots)$ with finite mean μ. Then, from (7.25), the total expected cost is

$$C_S(n; p_k) = c_S \sum_{k=1}^{\infty} p_k \int_0^{\infty} F^{(n)}(t)[1 - G^{(k)}(t)] \, dt$$

$$+ c_E \sum_{k=1}^{\infty} p_k \int_0^{\infty} [1 - F^{(n)}(t)]G^{(k)}(t) \, dt + c_0 n \quad (n = 1, 2, \ldots).$$

$$(7.49)$$

In particular, when $G(t) = 1 - e^{-\theta t}$ and $F(t) = 1 - e^{-\lambda t}$,

$$C_S(n; p_k) = \frac{c_S \mu}{\theta} + c_E \frac{n}{\lambda} + c_0 n$$

$$- \frac{c_S + c_E}{\theta + \lambda} \sum_{j=0}^{n-1} \sum_{i=0}^{\infty} \binom{i+j}{j} \left(\frac{\lambda}{\theta + \lambda}\right)^j \left(\frac{\theta}{\theta + \lambda}\right)^i \sum_{k=i+1}^{\infty} p_k.$$

From the inequality $C_S(n + 1; p_k) - C_S(n; p_k) \geq 0$ and (7.27),

$$\left(\frac{\lambda}{\theta + \lambda}\right)^{n+1} \sum_{j=0}^{\infty} \binom{n+j}{j} \left(\frac{\theta}{\theta + \lambda}\right)^j \sum_{k=j+1}^{\infty} p_k \leq \frac{c_E + \lambda c_0}{c_S + c_E}. \qquad (7.50)$$

Letting $L_S(n)$ be the left-hand side of (7.50),

Table 7.7 Optimum n^*, n_R^* and n_G^* when $c_E/c_S = 0.5$ and $\lambda c_0/c_S = 0.1$

λ/θ	$N = 2$			$N = 5$			$N = 10$		
	n^*	n_R^*	n_G^*	n^*	n_R^*	n_G^*	n^*	n_R^*	n_G^*
0.1	1	1	1	1	1	1	1	1	1
0.2	1	1	1	1	1	1	2	2	2
0.3	1	1	1	2	2	1	3	3	3
0.4	1	1	1	2	2	2	4	4	4
0.5	1	1	1	3	3	2	5	5	5
0.6	1	1	1	3	3	3	6	6	5
0.7	1	1	1	4	4	3	7	7	6
0.8	2	1	1	4	4	4	9	9	7
0.9	2	2	2	5	5	4	10	10	8

$$L_S(n) - L_S(n+1) = \left(\frac{\lambda}{\theta+\lambda}\right)^{n+1} \sum_{j=1}^{\infty} \binom{n+j}{j-1} \left(\frac{\theta}{\theta+\lambda}\right)^j p_j > 0.$$

Thus, by making similar arguments to Sect. 7.2.1, there exists a finite and unique minimum n_S^* $(1 \le n_S^* < \infty)$ which satisfies (7.50).

Example 7.7 (*Number for Poisson and geometric distributions*) When $p_k = [\beta^{k-1}/(k-1)!]e^{-\beta}$ $(k = 1, 2, \ldots)$ and $G(t) = 1 - e^{-\theta t}$, (7.50) is

$$\left(\frac{\lambda}{\theta+\lambda}\right)^{n+1} \sum_{j=0}^{\infty} \binom{n+j}{j} \left(\frac{\theta}{\theta+\lambda}\right)^j \sum_{k=j}^{\infty} \frac{\beta^k}{k!}e^{-\beta} \le \frac{c_E + \lambda c_0}{c_S + c_E}. \qquad (7.51)$$

When $p_k = pq^{k-1}$ $(k = 1, 2, \ldots)$,

$$\left(\frac{\lambda}{\theta+\lambda}\right)^{n+1} \sum_{j=0}^{\infty} \binom{n+j}{j} \left(\frac{q\theta}{\theta+\lambda}\right)^j \le \frac{c_E + \lambda c_0}{c_S + c_E},$$

i.e., (Problem 10 in Sect. 7.4)

$$\left(\frac{\lambda}{\lambda+p\theta}\right)^{n+1} \le \frac{c_E + \lambda c_0}{c_S + c_E}. \qquad (7.52)$$

Table 7.7 presents optimum n^* which satisfies (7.27), when $\beta + 1 = N$, $p = 1/N$, $c_E/c_S = 0.5$ and $\lambda c_0/c_S = 0.1$, n_R^* and n_G^* which satisfy (7.51) and (7.52), respectively. All n^*, n_R^* and n_G^* increase with N and λ/θ, and $n^* \ge n_R^*$ and n_G^*. In addition, both n^* and n_R^* are almost the same, and $n^*/\lambda \approx 10/\theta$. $\qquad \square$

7.3.4 Parallel System

Consider a parallel system with n units for a random number N of works. Then, by replacing $F^{(n)}(t)$ in (7.49) with $F(t)^n$, the total expected cost is

$$C_P(n; p_k) = c_S \sum_{k=1}^{\infty} p_k \int_0^{\infty} F(t)^n [1 - G^{(k)}(t)] \, dt$$

$$+ c_E \sum_{k=1}^{\infty} p_k \int_0^{\infty} [1 - F(t)^n] G^{(k)}(t) \, dt + c_0 n \quad (n = 1, 2, \ldots). \quad (7.53)$$

In particular, when $G(t) = 1 - e^{-\theta t}$ and $F(t) = 1 - e^{-\lambda t}$,

$$C_P(n; p_k) = \frac{c_S \mu}{\theta} + \frac{c_E}{\lambda} \sum_{j=1}^{n} \frac{1}{j} + c_0 n$$

$$+ \frac{c_S + c_E}{\lambda} \sum_{j=1}^{n} (-1)^j \binom{n}{j} \frac{1}{j} \sum_{k=1}^{\infty} p_k \left[1 - \left(\frac{\theta}{j\lambda + \theta} \right)^k \right]. \quad (7.54)$$

From the inequality $C_P(n + 1; p_k) - C_P(n; p_k) \geq 0$ and (7.31),

$$\sum_{j=0}^{n} (-1)^j \binom{n}{j} \frac{1}{j+1} \sum_{k=1}^{\infty} p_k \left\{ 1 - \left[\frac{\theta}{(j+1)\lambda + \theta} \right]^k \right\} \leq \frac{c_E/(n+1) + c_0 \lambda}{c_S + c_E}.$$

$$(7.55)$$

Letting $L_P(n)$ be the left-hand side of (7.55),

$$L_P(n) - L_P(n+1) = \sum_{j=0}^{n} (-1)^j \binom{n}{j} \frac{1}{j+2} \sum_{k=1}^{\infty} p_k \left\{ 1 - \left[\frac{\theta}{(j+2)\lambda + \theta} \right]^k \right\} > 0.$$

Thus, $L_P(n)$ decreases strictly with n to 0, and the right-hand side of (7.55) decreases to $c_0 \lambda / (c_S + c_E)$. Therefore, there exists a finite and unique minimum n_P^* ($1 \leq n_P^* < \infty$) which satisfies (7.55).

Example 7.8 (Number for Poisson and geometric distributions) When $p_k = [\beta^{k-1} / (k-1)!] e^{-\beta}$ $(k = 1, 2, \ldots)$ and $G(t) = 1 - e^{-\theta t}$, (7.55) is

Table 7.8 Optimum n^*, n_R^* and n_G^* when $c_E/c_S = 0.5$ and $\lambda c_0/c_S = 0.1$

λ/θ	$N = 2$			$N = 5$			$N = 10$		
	n^*	n_R^*	n_G^*	n^*	n_R^*	n_G^*	n^*	n_R^*	n_G^*
0.1	1	1	1	1	1	1	1	1	1
0.2	1	1	1	1	1	1	3	3	2
0.3	1	1	1	2	2	1	5	5	3
0.4	1	1	1	3	3	2	7	6	4
0.5	1	1	1	4	3	2	8	7	4
0.6	1	1	1	5	4	3	8	8	5
0.7	2	1	1	5	5	3	9	8	5
0.8	2	2	2	6	5	4	9	9	6
0.9	2	2	2	7	6	4	9	9	6

$$\sum_{j=0}^{n} (-1)^j \binom{n}{j} \frac{1}{j+1} \left\{ 1 - \frac{\theta}{(j+1)\lambda + \theta} \exp\left[-\beta \frac{(j+1)\lambda}{(j+1)\lambda + \theta} \right] \right\}$$
$$\leq \frac{c_E/(n+1) + c_0\lambda}{c_S + c_E}. \tag{7.56}$$

Next, when $p_k = pq^{k-1}$ $(k = 1, 2, \ldots)$, (7.55) is

$$\sum_{j=0}^{n} (-1)^j \binom{n}{j} \frac{\lambda}{(j+1)\lambda + \theta p} \leq \frac{c_E/(n+1) + c_0\lambda}{c_S + c_E}. \tag{7.57}$$

Table 7.8 presents optimum n^* which satisfies (7.31), when $\beta + 1 = N$, $p = 1/N$, $c_E/c_S = 0.5$ and $\lambda c_0/c_S = 0.1$, n_R^* and n_G^* which satisfy (7.56) and (7.57), respectively. This shows a similar tendency to Table 7.8. □

7.4 Problems

1. Derive (7.9) and compute Table 7.2.
2. Prove that $L_j = L^*$ for all $j \geq 1$ and L^* increases strictly with $1/\theta$ to ∞.
3. Compute γT_j^* and γL_j^* numerically from (7.9) using Algorithm [13, p. 112], [14, p. 203].
4. Compute numerically both optimum L_{T2}^*, N_T^* and L_{P2}^*, N_P^* which minimize $C_T(L, N)$ and $C_P(L, N)$, respectively.
5. Derive (7.22) and prove that L_{P2}^* increases strictly with N.
6. Derive (7.26) and (7.30).
*7. Discuss analytically both optimum n^* and N^* which minimize $C_S(n, N)$ in (7.26) and compute numerically them.

*8. Discuss analytically both n^* and N^* which minimize $C_P(n, N)$ in (7.30) and compute numerically them.
*9. Derive analytically and compute numerically optimum n^* and N^* which minimize $C_S(n, N)$ in (7.26) and $C_P(n, N)$ in (7.30), respectively.
10. Derive (7.52).

References

1. Dempster NAH, Lenstra JK (eds) (1982) Deterministic and stochastic scheduling. Reidel, Dordrecht
2. Forst FG (1984) A review of the static stochastic job sequencing literature. Oper Res 21:127–144
3. Righter R (1994) Stochastic scheduling. In: Shaked M, Shanthikumar G (eds) Stochastic orders. Academic Press, San Diego
4. Pinedo M (2008) Scheduling theory, algorithm and systems. Prentice Hall, NJ
5. Sarin SC, Erel E, Steiner G (1991) Sequencing jobs on a single machine with a common due date and stochastic processing times. Euro J Oper Res 51:188–198
6. Soroush HM, Fredendall LD (1994) The stochastic single machine scheduling problem with earliness and tardiness costs. Euro J Oper Res 77:287–302
7. Soroush HM (1996) Optimal sequences in stochastic single machine shops. Comput Oper Res 23:705–721
8. Pinedo M, Ross SM (1980) Scheduling jobs subject to nonhomogeneous Poisson shocks. Manage Sci 26:1250–1257
9. Zhou X, Cai X (1997) General stochastic single-machine scheduling with regular cost functions. Math Comput Model 26:95–108
10. Nakagawa T (2008) Advanced reliability models and maintenance policies. Springer, London
11. Chen M, Nakagawa T (2012) Optimal scheduling of random works with reliability applications. Asia Pac J Oper Res 29:1250027 (14 pages)
12. Chen M, Nakagawa T (2013) Optimal redundant systems for works with random processing time. Reliab Eng Syst Saf 116:99–104
13. Barlow RE, Proschan F (1965) Mathematical theory of reliability. Wiley, New York
14. Nakagawa T (2005) Maintenance theory of reliability. Springer, London
15. Ushakov IA (1994) Handbook of reliability engineering. Wiley, New York
16. Ross M (1983) Stochastic processes. Wiley, New York
17. Nakagawa T (2011) Stochastic processes with applications to reliability theory. Springer, London
18. Bhat UN, Basawa IV (1992) Queueing and related models. Oxford University Press, Oxford

Chapter 8
Other Random Maintenance Models

We finally propose the following four random reliability models:

(1) Random Finite Interval.
(2) Random Interval Reliability.
(3) Random Failure Level.
(4) Other Random Models.

The unit sometimes has to be operating for a finite interval S which is random, because the working times of a job might be random. Then, we take up inspection policies, replacement policies with minimal repair and imperfect preventive maintenance policies for a random interval. The expected costs of each policy are obtained, and optimum policies which minimize them are discussed analytically and numerically. Furthermore, we consider replacement policies with a discount rate for a random interval.

Interval reliability is defined as the probability that at time T_0, the unit is operating and will continue to operate for an interval x. We consider two cases where T_0 and x are random. When the preventive maintenance is done at time T, optimum policies which maximize the interval reliabilities are derived.

We have studied the replacement policies for cumulative damage models [1, p. 40], where the unit is replaced before failure at a specified number N of shocks when they occur at random times, and fails when the total damage exceeds a failure level K. We take up two replacement policies where K is random, because most units have individual variations in their ability to withstand shocks and are operating in a different environment [1, p. 29]. Two expected cost rates are obtained and optimum policies which minimize them are derived. Finally, we propose three random reliability models about random scheduling time in Chap. 7, random inspection number in Chap. 4 and periodic replacement with random number of failures in Chap. 3. Optimum policies which minimize the expected costs of each model are derived. Such modified and extended models would give interesting topics for further studies in maintenance and reliability theory.

© Springer-Verlag London 2014
T. Nakagawa, *Random Maintenance Policies*,
Springer Series in Reliability Engineering, DOI 10.1007/978-1-4471-6575-0_8

8.1 Random Finite Interval

This section takes up some maintenance models in which the working interval of units
is uncertain and is given as a random variable. We obtain the total expected costs for a
finite interval and discuss analytically optimum policies which minimize them, using
the known results of maintenance policies. However, there have been little papers
treated with maintenance models for a finite interval, because it is more difficult
theoretically to discuss optimum policies. The optimum replacement policies with
random life cycle were discussed analytically, using a discount rate [2]. Inspection
policies and modified models for a finite interval were discussed as partition problems
[3, p. 39], [4], and periodic and sequential inspection policies for a finite interval were
summarized [5]. Some maintenance models with the expected present value of total
cost under a random life cycle were considered [6].

We summarize some optimum policies in which the unit works for a random finite
interval S $(0 \le S < \infty)$: A finite interval S has a general distribution $G(s) \equiv \Pr\{S \le
s\}$ with finite mean $1/\theta$ $(0 < \theta < \infty)$. The failure time of the unit has a probability
distribution $F(t)$ with finite mean μ $(0 < \mu < \infty)$ and a density function $f(t)$, i.e.,
$\mu \equiv \int_0^\infty \overline{F}(t)\, dt$ and $f(t) \equiv dF(t)/dt$, where $\overline{\Phi}(t) \equiv 1 - \Phi(t)$ for any function
$\Phi(t)$. It is assumed that the failure rate $h(t) \equiv f(t)/\overline{F}(t)$ increases with t to $h(\infty)$,
and the cumulative hazard rate is $H(t) \equiv \int_0^t h(u)\, du$, i.e., $F(t) = 1 - e^{-H(t)}$.

Under the above assumptions, we propose the following three maintenance models
for a random finite interval:

(1) Periodic and sequential inspections.
(2) Periodic and sequential replacements with minimal repair.
(3) Imperfect PM (Preventive Maintenance).

When the failure time is exponential, i.e., $F(t) = 1 - e^{-\lambda t}$ $(0 < \lambda < \infty)$, we
derive optimum policies analytically which minimize the expected costs and compute
optimum times numerically.

8.1.1 Inspection Policies

8.1.1.1 Periodic Inspection

An operating unit is checked at periodic time kT $(k = 1, 2, \ldots)$ $(0 < T \le \infty)$ and
is replaced at failure detection or at time S, whichever occurs first. Let c_T be the cost
for one check, c_D be the downtime cost per unit of time for the time elapsed between
a failure and its detection or S, and c_R be the cost for replacement. The expected cost
for a finite interval was obtained [3, p. 65], [5]. The total expected cost for a random
interval S is classified in the following three cases:

(a) When the unit fails at time t $(t < S)$ and the next check occurs before time S,
    the expected cost is (Problem 1 in Sect. 8.5)

$$\sum_{k=0}^{\infty} \overline{G}[(k+1)T] \int_{kT}^{(k+1)T} \{kc_T + c_R + c_D[(k+1)T - t]\} \, dF(t).$$

(b) When the unit fails at time t $(t < S)$ and the next inspection occurs after time S, the expected cost is

$$\sum_{k=0}^{\infty} \int_{kT}^{(k+1)T} \left\{ \int_{t}^{(k+1)T} [kc_T + c_R + c_D(u - t)] \, dG(u) \right\} dF(t).$$

(c) When the unit dose not fail until time S, the expected cost is

$$\sum_{k=0}^{\infty} (kc_T + c_R) \int_{kT}^{(k+1)T} \overline{F}(t) \, dG(t).$$

Summing up (a)–(c), the total expected cost until replacement is (Problem 2 in Sect. 8.5)

$$C_1(T) = c_R + c_T \sum_{k=0}^{\infty} k \left[\int_{kT}^{(k+1)T} \overline{G}(t) \, dF(t) + \int_{kT}^{(k+1)T} \overline{F}(t) \, dG(t) \right]$$

$$+ c_D \sum_{k=0}^{\infty} \int_{kT}^{(k+1)T} \left[\int_{t}^{(k+1)T} \overline{G}(u) \, du \right] dF(t)$$

$$= c_R - c_T + \sum_{k=0}^{\infty} \overline{F}(kT) \left[c_T \overline{G}(kT) + c_D \int_{kT}^{(k+1)T} \overline{G}(t) \, dt \right]$$

$$- c_D \int_{0}^{\infty} \overline{F}(t) \overline{G}(t) \, dt. \tag{8.1}$$

Clearly,

$$C_1(0) \equiv \lim_{T \to 0} C_1(T) = \infty,$$

$$C_1(\infty) \equiv \lim_{T \to \infty} C_1(T) = c_R + c_D \int_{0}^{\infty} F(t) \overline{G}(t) \, dt.$$

Thus, there exists a positive T^* $(0 < T^* \le \infty)$ which minimizes (8.1).

We find an optimum T^* which minimizes

$$\tilde{C}_1(T) = C_1(T) - c_R + c_T + c_D \int_0^\infty \overline{F}(t)\overline{G}(t)\,dt.$$

In particular, when $F(t) = 1 - e^{-\lambda t}$ $(0 < \lambda < \infty)$ and $G(t) = 1 - e^{-\theta t}$ $(0 < \theta < \infty)$,

$$\tilde{C}_1(T) = \frac{c_T + (c_D/\theta)(1 - e^{-\theta T})}{1 - e^{-(\theta+\lambda)T}}, \tag{8.2}$$

which agrees with (4.18) in Chap. 4 when $\theta \to 0$. Differentiating $\tilde{C}_1(T)$ with respect to T and setting it to zero,

$$\frac{e^{\lambda T} - e^{-\theta T}}{\theta + \lambda} - \frac{1 - e^{-\theta T}}{\theta} = \frac{c_T}{c_D}, \tag{8.3}$$

whose left-hand side increases strictly with T from 0 to ∞. Thus, there exists a finite and unique T^* $(0 < T^* < \infty)$ which satisfies (8.3). It can be easily seen that the left-hand side of (8.3) decreases with θ to 0, and hence, T^* decreases with $1/\theta$ (Problem 3 in Sect. 8.5). When $1/\theta \to \infty$, i.e., the mean working time is infinite, (8.3) becomes

$$\frac{e^{\lambda T} - (1 + \lambda T)}{\lambda} = \frac{c_T}{c_D},$$

which agrees with the standard inspection policy [8, p. 204] with optimum time T_S^* given in (4.19). This means that T^* decreases with $1/\theta$ to T_S^*.

Example 8.1 (Checking time for exponential failure and interval times) Table 8.1 presents optimum T^* for $1/\theta$ and c_T/c_D when $F(t) = 1 - e^{-t/100}$ and $G(t) = 1 - e^{-\theta t}$. This indicates that T^* increases with c_T/c_D and decreases slowly with $1/\theta$ to T_S^*. It is of interest that if the checking cost c_T becomes four times, then T^* becomes almost two times (Problem 4 in Sect. 8.5). □

8.1.1.2 Sequential Interval

An operating unit is checked at successive times T_k $(k = 1, 2, \ldots)$, where $T_0 \equiv 0$. The total expected cost is, by replacing kT with T_k in (8.1),

Table 8.1 Optimum T^* when $F(t) = 1 - e^{-t/100}$ and $G(t) = 1 - e^{-\theta t}$

c_T/c_D	$1/\theta$					
	50	100	200	500	1,000	∞
0.5	10.161	9.996	9.915	9.868	9.852	9.836
1.0	14.458	14.130	13.972	13.878	13.847	13.817
2.0	20.615	19.967	19.656	19.475	19.415	19.355
3.0	25.397	24.434	23.977	23.709	23.622	23.534
4.0	29.463	28.191	27.590	27.240	27.125	27.011
5.0	33.069	31.492	30.751	30.321	30.181	30.040

$$C_1(T_1, T_2, \ldots) = c_R - c_T + \sum_{k=0}^{\infty} \overline{F}(T_k) \left[c_T \overline{G}(T_k) + c_D \int_{T_k}^{T_{k+1}} \overline{G}(t)\, dt \right]$$

$$- c_D \int_0^{\infty} \overline{F}(t)\overline{G}(t)\, dt. \tag{8.4}$$

When $G(t) = 1 - e^{-\theta t}$,

$$C_1(T_1, T_2, \ldots) = c_R - c_T + \sum_{k=0}^{\infty} \overline{F}(T_k) \left[\left(c_T + \frac{c_D}{\theta} \right) e^{-\theta T_k} - \frac{c_D}{\theta} e^{-\theta T_{k+1}} \right]$$

$$- c_D \int_0^{\infty} \overline{F}(t) e^{-\theta t}\, dt. \tag{8.5}$$

Differentiating $C_1(T_1, T_2, \ldots)$ with respect to T_k and setting it equal to zero,

$$\frac{1 - e^{-\theta(T_{k+1} - T_k)}}{\theta} = \frac{F(T_k) - F(T_{k-1})}{f(T_k)} - \frac{c_T}{c_D} \left[1 + \frac{\theta \overline{F}(T_k)}{f(T_k)} \right]. \tag{8.6}$$

Using the algorithm [7, p. 112], [8, p. 203], we compute optimum sequence times $\{T_k^*\}$ which satisfy (8.6). When $1/\theta = \infty$, (8.6) becomes

$$T_{k+1} - T_k = \frac{F(T_k) - F(T_{k-1})}{f(T_k)} - \frac{c_T}{c_D},$$

which agrees with (4.17) and corresponds to the standard sequential inspection policy [7, p. 110], [8, p. 203].

Example 8.2 (Sequential time for exponential interval and Weibull failure times)
Table 8.2 presents optimum T_k^* ($k = 1, 2, \ldots, 10$) for $1/\theta$ when $F(t) = 1 - \exp[-(\lambda t)^2]$, $1/\lambda = 100$, $\mu = 88.6$, and $c_T/c_D = 5$. For example, when $1/\theta = 200$,

Table 8.2 Optimum T_k^* when $F(t) = 1 - e^{-(t/100)^2}$, $G(t) = 1 - e^{-\theta t}$ and $c_T/c_D = 5$

k	\multicolumn{5}{c}{$1/\theta$}
	100	200	500	1,000	∞
1	56.90	55.96	55.39	55.20	55.01
2	84.60	83.37	82.63	82.38	82.12
3	107.43	106.01	105.14	104.85	104.55
4	127.62	126.04	125.07	124.74	124.74
5	146.05	144.34	143.28	142.92	142.55
6	163.19	161.34	160.20	159.80	159.40
7	179.30	177.32	176.08	175.64	175.20
8	194.55	192.41	191.05	190.58	190.09
9	209.01	206.68	205.18	204.65	204.10
10	222.65	220.07	218.38	217.78	217.16

T_2^* is almost the same mean failure time μ and T_9^* is almost the same mean time interval $1/\theta$. This indicates that T_k^* decreases slowly with $1/\theta$ to optimum times of periodic inspection, and the differences between T_k^* and T_{k+1}^* also decrease with k and $1/\theta$.   □

8.1.2 Replacement with Minimal Repair

8.1.2.1 Periodic Replacement

An operating unit is replaced at periodic times kT ($k = 1, 2, \ldots$) and any unit is as good as new one at each replacement in Fig. 8.1. When the unit fails, only minimal repair is made, its failure rate remains undisturbed by any repair of failures [7, p. 96], [8, p. 96]. It is assumed that the repair and replacement times are negligible and the failure rate $h(t)$ increases strictly to $h(\infty)$. Suppose that the unit works for a random finite interval S. Let c_M be the cost for minimal repair. The other assumptions are the same as those of Sect. 8.1.1. Then, the total expected cost until replacement is

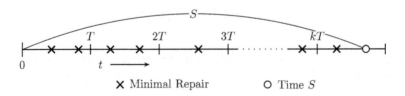

Fig. 8.1 Periodic replacement with minimal repair

$$C_2(T) = \sum_{k=0}^{\infty} \int_{kT}^{(k+1)T} \{k[c_T + c_M H(T)] + c_M H(t - kT) + c_R\} \, dG(t)$$

$$= c_R - c_T + c_T \sum_{k=0}^{\infty} \overline{G}(kT) + c_M \sum_{k=0}^{\infty} \int_0^T \overline{G}(t + kT) h(t) \, dt. \qquad (8.7)$$

In particular, when $G(t) = 1 - e^{-\theta t}$ (Problem 5 in Sect. 8.5),

$$C_2(T) = c_R - c_T + \frac{c_T + c_M \int_0^T e^{-\theta t} h(t) \, dt}{1 - e^{-\theta T}}. \qquad (8.8)$$

Differentiating $C_2(T)$ with respect to T and setting it equal to zero,

$$\int_0^T e^{-\theta t} [h(T) - h(t)] \, dt = \int_0^T \frac{1 - e^{-\theta t}}{\theta} \, dh(t) = \frac{c_T}{c_M}, \qquad (8.9)$$

whose left-hand side $L_1(T)$ increases strictly with T from 0 to

$$L_1(\infty) \equiv \int_0^{\infty} e^{-\theta t} [h(\infty) - h(t)] \, dt = \int_0^{\infty} \frac{1 - e^{-\theta t}}{\theta} \, dh(t).$$

Therefore, we have the following optimum policy:

(i) If $L_1(\infty) > c_T/c_M$, then there exists a finite and unique T^* $(0 < T^* < \infty)$ which satisfies (8.9), and the resulting cost rate is

$$C_2(T^*) = c_R - c_T + \frac{c_M}{\theta} h(T^*).$$

(ii) If $L_1(\infty) \le c_T/c_M$, then $T^* = \infty$, and

$$C_2(\infty) = c_R + c_M \int_0^{\infty} e^{-\theta t} h(t) \, dt.$$

Furthermore, when $1/\theta = \infty$, (8.9) becomes

$$\int_0^T t \, dh(t) = \frac{c_T}{c_M},$$

Table 8.3 Optimum T^* when $F(t) = 1 - e^{-(t/5)^2}$ and $G(t) = 1 - e^{-\theta t}$

c_T/c_M	$1/\theta$					
	50	100	200	500	1,000	∞
0.5	3.578	3.556	3.546	3.540	3.538	3.536
1.0	5.085	5.042	5.021	5.008	5.004	5.000
2.0	7.242	7.155	7.113	7.088	7.079	7.071
3.0	8.918	8.787	8.723	8.685	8.673	8.660
4.0	10.345	10.169	10.084	10.033	10.017	10.000
5.0	11.613	11.393	11.285	11.222	11.201	11.180

which agrees with (3.4) and corresponds to the standard periodic replacement [7, p. 97],[8, p. 102] with optimum time T_S^*. Thus, an optimum T^* decreases with $1/\theta$ to T_S^* because $L_1(T)$ increases with $1/\theta$.

Example 8.3 (Replacement for exponential interval and Weibull failure times) Table 8.3 presents optimum T^* for $1/\theta$ and c_T/c_M when $F(t) = 1 - \exp[-(\lambda t)^2]$, $1/\lambda = 5.0$ and $G(t) = 1 - e^{-\theta t}$. In this case, T^* is given by the solution of the equation

$$2 \left(\frac{\lambda}{\theta} \right)^2 [\theta T - (1 - e^{-\theta T})] = \frac{c_T}{c_M},$$

and when $1/\theta = \infty$,

$$T_S^* = \frac{1}{\lambda} \sqrt{\frac{c_T}{c_M}},$$

which shows that T^* decreases with $1/\theta$ to T_S^* (Problem 6 in Sect. 8.5). This indicates that optimum T^* has a similar tendency to Table 8.1. In addition, when $1/\theta = \infty$, if c_T becomes 4 times, then T_S^* becomes exactly two times.   □

8.1.2.2 Sequential Interval

An operating unit is replaced at successive times T_k ($k = 1, 2, \ldots$), where $T_0 \equiv 0$. The total expected cost is, by replacing kT with T_k in (8.7),

$$C_2(T_1, T_2, \ldots) = c_R - c_T + c_T \sum_{k=0}^{\infty} \overline{G}(T_k) + c_M \sum_{k=0}^{\infty} \int_0^{T_{k+1}-T_k} \overline{G}(t + T_k) h(t) \, dt.$$

$$(8.10)$$

In particular, when $G(t) = 1 - e^{-\theta t}$,

$$C_2(T_1, T_2, \ldots) = c_R - c_T + \sum_{k=0}^{\infty} e^{-\theta T_k} \left[c_T + c_M \int_0^{T_{k+1}-T_k} e^{-\theta t} h(t) \, dt \right]. \quad (8.11)$$

Differentiating $C_2(T_1, T_2, \ldots)$ with respect to T_k and setting it equal to zero,

$$\int_0^{T_{k+1}-T_k} e^{-\theta t} \, dh(t) = h(T_k - T_{k-1}) - \frac{\theta c_T}{c_M}. \quad (8.12)$$

Using the algorithm [7, p. 112], [8, p. 203], we compute optimum sequence times $\{T_k^*\}$ which satisfy (8.12). It can be seen that when $1/\theta = \infty$, $T_k^* = kT^*$ ($k = 1, 2, \ldots$).

Example 8.4 (Sequential time for exponential interval and Weibull failure times) When $F(t) = 1 - \exp[-(\lambda t)^2]$, (8.12) becomes

$$1 - e^{-\theta(T_{k+1}-T_k)} = \theta(T_k - T_{k-1}) - \frac{c_T}{2c_M} \left(\frac{\theta}{\lambda} \right)^2.$$

Table 8.4 presents optimum T_k^* ($k = 1, 2, \ldots, 10$) for $1/\theta$ when $1/\lambda = 5.0$ and c_P/c_M. This indicates that optimum T_k^* decreases with $1/\theta$ to $0.34k$. □

Table 8.4 Optimum T_k^* when $F(t) = 1 - e^{-(t/5)^2}$, $G(t) = 1 - e^{-\theta t}$ and $c_T/c_M = 5.0$

k	1/θ				
	100	200	500	1,000	∞
1	5.80	3.10	1.30	0.70	0.34
2	11.11	5.91	2.48	1.34	0.68
3	15.92	8.42	3.53	1.91	1.02
4	20.18	10.63	4.46	2.43	1.36
5	23.89	12.53	5.26	2.88	1.70
6	27.03	14.13	5.94	3.26	2.04
7	29.57	15.43	6.50	3.59	2.38
8	31.50	16.41	6.93	3.85	2.72
9	32.82	17.08	7.23	4.05	3.06
10	33.51	17.44	7.41	4.19	3.40

8.1.3 Imperfect PM

An operating unit undergoes imperfect PM [8, p. 171], [9] at successive times T_k ($k = 1, 2, \ldots$) for a random finite interval S, where $T_0 \equiv 0$: The failure rate between T_{k-1} and T_k is $B_k h(T_k - T_{k-1})$, where $1 = B_1 \le B_2 \le \cdots \le B_k \le \cdots$. When the unit fails between PMs, only minimal repair is made, i.e., the expected number of failures between T_{k-1} and T_k becomes $B_k H(T_k - T_{k-1})$. Then, the total expected cost until replacement is

$$
C_3(T_1, T_2, \ldots) = \sum_{k=0}^{\infty} \int_{T_k}^{T_{k+1}} \left\{ k c_T + c_R \right.
$$

$$
\left. + c_M \left[\sum_{j=1}^{k} B_j H(T_j - T_{j-1}) + B_{k+1}(t - T_k) \right] \right\} dG(t)
$$

$$
= c_R - c_T + c_T \sum_{k=0}^{\infty} \overline{G}(T_k)
$$

$$
+ c_M \sum_{k=0}^{\infty} B_{k+1} \int_0^{T_{k+1}-T_k} \overline{G}(t + T_k) h(t)\, dt, \tag{8.13}
$$

which agrees with (8.10) when $B_{k+1} \equiv 1$. In particular, when $G(t) = 1 - e^{-\theta t}$,

$$
C_3(T_1, T_2, \ldots) = c_R - c_T + c_T \sum_{k=0}^{\infty} e^{-\theta T_k}
$$

$$
+ c_M \sum_{k=0}^{\infty} B_{k+1} e^{-\theta T_k} \int_0^{T_{k+1}-T_k} e^{-\theta t} h(t)\, dt. \tag{8.14}
$$

Differentiating $C_3(T_1, T_2, \ldots)$ with respect to T_k and setting it equal to zero,

$$
B_{k+1} \int_0^{T_{k+1}-T_k} e^{-\theta t}\, dh(t) = B_k h(T_k - T_{k-1}) - \frac{\theta c_T}{c_M}. \tag{8.15}
$$

When $1/\theta = \infty$, (8.15) corresponds to the sequential imperfect PM with $N = \infty$ [8, p. 194].

Example 8.5 (Sequential time for exponential interval and Weibull failure times) It is assumed that $H(t) = (\lambda t)^2$, i.e., $h(t) = 2\lambda^2 t$. Then, (8.15) becomes

Table 8.5 Optimum T_k^* when $F(t) = 1 - e^{-(t/5)^2}$, $G(t) = 1 - e^{-\theta t}$ and $c_T/c_M = 5.0$

k	1/θ			
	100	200	500	∞
1	4.04	2.07	0.84	0.50
2	6.80	3.47	1.41	0.87
3	8.77	4.47	1.81	1.16
4	10.21	5.20	2.11	1.40
5	11.28	5.73	2.33	1.60
6	12.06	6.13	2.49	1.78
7	12.62	6.41	2.60	1.93
8	12.99	6.59	2.68	2.07
9	13.22	6.71	2.72	2.19
10	13.33	6.76	2.75	2.30

$$[1 - e^{-\theta(T_{k+1}-T_k)}]B_{k+1} = \theta(T_k - T_{k-1})B_k - \frac{c_T}{2c_M}\left(\frac{\theta}{\lambda}\right)^2.$$

Table 8.5 presents optimum $\{T_k^*\}$ when $1/\lambda = 5.0$, $c_P/c_M = 5$, and $B_{k+1} = (1 + 1/2)(1 + 1/3) \cdots (1 + 1/(k+1))$ ($k = 1, 2, \ldots$) (Problem 7 in Sect. 8.5), where $B_1 = 1$. This indicates that T_k^* decreases with $1/\theta$ and $T_{k+1}^* - T_k^*$ decreases with k.

□

8.1.4 Random Interval with Discount Rate

8.1.4.1 Periodic Replacement

When we adopt the total expected cost as an appropriate objective function, we should evaluate the present values of any maintenance costs by using an appropriate discount rate. Then, when a discount cost is α ($0 < \alpha < \infty$), the total expected costs of age and periodic replacements were obtained and their optimum policies were derived analytically [8, p. 78, p. 107, p. 119].

We consider periodic replacement with a continuous discount rate α for a random finite interval introduced in Sect. 8.1.2: An operating unit is replaced at periodic times kT ($k = 1, 2, \ldots$). Then, it is assumed that the present values of cost c at time t is $c\,e^{-\alpha t}$ at time 0. In this case, the present value of the total cost for replacement and minimal repair in the interval $[(k-1)T, kT]$ is

$$C_k = c_T + c_M \int_0^T e^{-\theta t} h(t)\, dt \quad (k = 1, 2, \ldots),$$

and hence, the present value at time 0 is given by $C_k e^{-\alpha kT}$. Therefore, from (8.7), the total expected cost with a discount rate α is

$$
\begin{aligned}
C(T;\alpha) = \sum_{k=0}^{\infty} \int_{kT}^{(k+1)T} & \left\{ \sum_{j=0}^{k-1} \left[c_T + c_M \int_0^T e^{-\alpha t} h(t)\, dt \right] e^{-\alpha jT} \right. \\
& \left. + \left[c_M \int_0^{t-kT} e^{-\alpha u} h(u)\, du + c_R \right] e^{-\alpha kT} \right\} dG(t),
\end{aligned}
\tag{8.16}
$$

where $\sum_{j=0}^{-1} \equiv 0$. In particular, when $G(t) = 1 - e^{-\theta t}$,

$$
C(T;\alpha) = \frac{c_R(1 - e^{-\theta T}) + c_T e^{-\theta T} + c_M \int_0^T e^{-(\theta+\alpha)t} h(t)\, dt}{1 - e^{-(\theta+\alpha)T}},
\tag{8.17}
$$

which agrees with (8.8) when $\alpha = 0$. Differentiating $C(T;\alpha)$ with respect to T and setting it equal to zero,

$$
c_M \int_0^T [1 - e^{-(\theta+\alpha)t}]\, dh(t)
$$

$$
+ (c_R - c_T)[\theta(e^{\alpha T} - 1) - \alpha(1 - e^{-\theta T})] = c_T(\theta + \alpha).
\tag{8.18}
$$

If $c_R \geq c_T$, then the left-hand side of (8.18) increases strictly from 0. For example, when $h(\infty) = \infty$ and $c_R \geq c_T$, there exists a finite and unique T^* ($0 < T^* < \infty$) which satisfies (8.18). When $c_R = c_T$, T^* increases with both θ and α. In the case of $F(t) = 1 - e^{-\lambda t}$ and $c_R > c_T$, (8.18) becomes

$$
\frac{\theta(e^{\alpha T} - 1) - \alpha(1 - e^{-\theta T})}{\theta + \alpha} = \frac{c_T}{c_R - c_T}.
$$

Thus, a finite T^* always exists uniquely.

8.1.4.2 Sequential Interval

An operating unit is replaced at successive times T_k ($k = 1, 2, \ldots$). Then, the total expected cost is, by replacing kT with T_k in (8.16),

$$C(T_1, T_2, \ldots; \alpha) = \sum_{k=0}^{\infty} \int_{T_k}^{T_{k+1}} \left\{ \sum_{j=0}^{k-1} \left[c_T + c_M \int_0^{T_{j+1}-T_j} e^{-\alpha t} h(t) \, dt \right] e^{-\alpha T_j} \right.$$

$$+ \left. \left[c_M \int_0^{t-T_k} e^{-\alpha u} h(u) \, du + c_R \right] e^{-\alpha T_k} \right\} dG(t). \qquad (8.19)$$

In particular, when $G(t) = 1 - e^{-\theta t}$,

$$C(T_1, T_2, \ldots; \alpha) = c_R \sum_{k=0}^{\infty} e^{-(\theta+\alpha)T_k} - (c_R - c_T) \sum_{k=0}^{\infty} e^{-\theta T_{k+1} - \alpha T_k}$$

$$+ c_M \sum_{k=0}^{\infty} e^{-(\theta+\alpha)T_k} \int_0^{T_{k+1}-T_k} e^{-(\theta+\alpha)t} h(t) \, dt, \qquad (8.20)$$

which agrees with (8.11) when $\alpha = 0$. Differentiating $C(T_1, T_2, \ldots; \alpha)$ with respect to T_k and setting it equal to zero (Problem 8 in Sect. 8.5),

$$c_M \int_0^{T_{k+1}-T_k} e^{-(\theta+\alpha)t} \, dh(t) + (c_R - c_T)\alpha[1 - e^{-\theta(T_{k+1}-T_k)}]$$

$$= c_M h(T_k - T_{k-1}) + (c_R - c_T)\theta[e^{\alpha(T_k - T_{k-1})} - 1] - c_T(\theta + \alpha), \qquad (8.21)$$

which agrees with (8.12) when $\alpha = 0$ and (8.18) when $T_k = kT$.

8.2 Random Interval Reliability

Interval reliability $R(x, T_0)$ is defined as the probability that at a specified time T_0 $(0 \le T_0 < \infty)$, the unit is operating and will continue to operate for an interval of time x $(0 \le x < \infty)$ [7, p. 74] [8, p. 48]. A typical model is a standby generator in which T_0 is the time until the electric power stops and x is the required time until the electric power recovers. In this case, the interval reliability represents the probability that a standby generator will be able to operate while the electric power is interrupted.

Consider a standard one-unit system which is repaired upon failure and is brought back to operation after the repair completion [8, p. 40] in Examples 1.2 and 1.4: The failure time has a general distribution $F(t)$ with finite mean μ $(0 < \mu < \infty)$ and the repair time has a general distribution $G(t)$ with finite mean β $(0 < \beta < \infty)$. We suppose that the unit is in State 0 when it is operating and begins to operate at time 0. Let $M_{00}(t)$ be the expected number of visiting State 0 during $(0, t]$. Then, the Laplace–Stieltjes (LS) transform of $M_{00}(t)$ is [8, p. 41]

$$M_{00}^*(s) = \int_0^\infty e^{-st}\, dM_{00}(t) = \frac{F^*(s)G^*(s)}{1 - F^*(s)G^*(s)},$$

where $\Phi^*(s) \equiv \int_0^\infty e^{-st}\, d\Phi(t)$ for $\mathrm{Re}(s) > 0$ for any function $\Phi(t)$. Thus, the interval reliability is [8, p. 48]

$$R(x, T_0) = \overline{F}(T_0 + x) + \int_0^T \overline{F}(T_0 + x - t)\, dM_{00}(t),$$

and its Laplace transform is

$$R^*(x, s) \equiv \int_0^\infty e^{-sT_0} R(x, T_0)\, dT_0 = \frac{\int_0^\infty e^{-st}\overline{F}(t + x)\, dt}{1 - F^*(s)G^*(s)}. \tag{8.22}$$

Thus, the limiting interval reliability is

$$R(x) \equiv \lim_{T_0 \to \infty} R(x, T_0) = \lim_{s \to 0} s R^*(x, s) = \frac{\int_0^\infty \overline{F}(t + x)\, dt}{\mu + \beta}. \tag{8.23}$$

Next, we set the PM time T ($0 < T \le \infty$) for the operating unit. However, the PM of the operating unit is not done during the interval even if it is the time for PM. It is assumed that the distribution of time for PM is the same as the repair distribution $G(t)$. For the unit with PM time T, the LS transform of $M_{00}(t)$ which is the expected number of visiting State 0 is [8, p. 138]

$$M_{00}^*(s) = \frac{G^*(s)[1 - \int_0^T \overline{F}(t)se^{-st}\, dt]}{1 - G^*(s)[1 - \int_0^T \overline{F}(t)se^{-st}\, dt]}.$$

In a similar of obtaining (8.22), the interval reliability is

$$R(T; x, T_0) = \overline{F}(T_0 + x)\overline{D}(T_0) + \int_0^{T_0} \overline{F}(T_0 + x - t)\overline{D}(T_0 - t)\, dM_{00}(t),$$

where $D(t)$ is the degenerating distribution placing unit mass at T, i.e., $D(t) \equiv 0$ for $t < T$ and 1 for $t \ge T$, and its Laplace transform is

$$R^*(T; x, s) \equiv \int_0^\infty e^{-sT_0} R(T; x, T_0) \, dT_0$$

$$= \frac{\int_0^T e^{-st} \overline{F}(t + x) \, dt}{1 - G^*(s) + G^*(s) \int_0^T \overline{F}(t) s e^{-st} \, dt}. \tag{8.24}$$

Thus, the limiting interval reliability is

$$R(T; x) \equiv \lim_{T_0 \to \infty} R(T; x, T_0) = \lim_{s \to 0} s R^*(T; x, s)$$

$$= \frac{\int_0^T \overline{F}(t + x) \, dt}{\int_0^T \overline{F}(t) \, dt + \beta}, \tag{8.25}$$

which agrees with (8.23) when $T = \infty$.

8.2.1 Random Time

When T_0 is a random variable with an exponential distribution $(1 - e^{-\theta t})$ $(0 < \theta < \infty)$, the interval reliability is, from (8.24),

$$R(T; x, \theta) \equiv \int_0^\infty R(T; x, T_0) \, d(1 - e^{-\theta T_0})$$

$$= \frac{\int_0^T \overline{F}(t + x) \theta e^{-\theta t} \, dt}{1 - G^*(\theta) + G^*(\theta) \int_0^T \overline{F}(t) \theta e^{-\theta t} \, dt}, \tag{8.26}$$

which agrees with $R(T; x)$ in (8.25) as $\theta \to 0$.

We find an optimum PM time T_1^* which maximizes the interval reliability $R(T; x, \theta)$ for a fixed $x > 0$. Let $\lambda(t; x) \equiv [F(t + x) - F(t)]/\overline{F}(t)$ for $t \geq 0$, which is the probability that the unit with age t fails in an interval $(t, t + x]$ and is called the same failure rate as $h(t)$, because they have the same property [7, p. 23], [8, p. 6]. It is assumed that $\lambda(t; x)$ increases strictly with t from $F(x)$ to $\lambda(\infty; x) \leq 1$. Differentiating $R(T; x, \theta)$ with respect to T and setting it equal to zero,

$$\lambda(T; x) \left[G^*(\theta) \int_0^T \overline{F}(t) \theta e^{-\theta t} \, dt + 1 - G^*(\theta) \right]$$

$$+ \int_0^T [\overline{F}(t + x) - G^*(\theta) \overline{F}(t)] \theta e^{-\theta t} \, dt = 1 - G^*(\theta), \tag{8.27}$$

whose left-hand increases strictly from $F(x)[1 - G^*(\theta)]$ to

$$\lambda(\infty; x)[1 - F^*(\theta)G^*(\theta)] + \int_0^\infty \overline{F}(t + x)\theta e^{-\theta t}\, dt - G^*(\theta)[1 - F^*(\theta)].$$

Therefore, if $\lambda(\infty; x) > K_1(\theta)$, then there exists a finite and unique T_1^* ($0 < T_1^* < \infty$) which satisfies (8.27), and the resulting interval reliability is

$$R(T_1^*; x, \theta) = \frac{1 - \lambda(T_1^*; x)}{G^*(\theta)}. \tag{8.28}$$

If $\lambda(\infty; x) \le K_1(\theta)$, then $T_1^* = \infty$, and the interval reliability is

$$R(\infty; , x, \theta) = \frac{\int_0^\infty \overline{F}(t + x)\theta e^{-\theta t}\, dt}{1 - F^*(\theta)G^*(\theta)} = \theta R^*(x, \theta), \tag{8.29}$$

where

$$K_1(\theta) \equiv 1 - R(\infty; x, \theta).$$

Note that optimum T_1^* goes to ∞ as x becomes larger, i.e., $\lambda(t; x) \to 1$ for $t \ge 0$, because the left-hand side of (8.27) goes to $1 - G^*(\theta)$ as $x \to \infty$. When x also goes to 0, $R(T; 0, \theta)$ increases with T, and hence, $T_1^* \to \infty$.

Next, suppose that X of the limiting interval reliability $R(T; X)$ in (8.25) is a random variable with an exponential distribution $\Pr\{X \le x\} = 1 - e^{-\theta x}$. Then, the limiting interval reliability is, from (8.25),

$$R(T; \theta) \equiv \int_0^\infty R(T; x)\, d(1 - e^{-\theta x}) = \frac{\int_0^\infty [\int_0^T \overline{F}(t + x)\, dt]\theta e^{-\theta x}\, dx}{\int_0^T \overline{F}(t)\, dt + \beta}. \tag{8.30}$$

We find an optimum PM time T_2^* which maximizes $R(T; \theta)$ for a fixed θ. Differentiating $R(T; \theta)$ with respect to T and setting it equal to zero,

$$\int_0^\infty \theta e^{-\theta x}\lambda(T; x)\, dx \left[\int_0^T \overline{F}(t)\, dt + \beta\right]$$
$$- \int_0^\infty \theta e^{-\theta x}\left\{\int_0^T [F(t + x) - F(t)]\, dt\right\} dx = \beta, \tag{8.31}$$

whose left-hand increases from $\beta \int_0^\infty \theta e^{-\theta t} F(t)\, dt$ to

$$(\mu + \beta) \int_0^\infty \theta e^{-\theta t} \lambda(\infty; t)\, dt - \int_0^\infty \overline{F}(t) e^{-\theta t}\, dt.$$

Therefore, if

$$\int_0^\infty \overline{F}(t)(1 - e^{-\theta t})\, dt > (\mu + \beta) \int_0^\infty \theta e^{-\theta t}[1 - \lambda(\infty; t)]\, dt,$$

then there exists a finite and unique T_2^* $(0 < T_2^* < \infty)$ which satisfies (8.31), and the resulting reliability is

$$R(T_2^*; \theta) = 1 - \int_0^\infty \theta e^{-\theta t} \lambda(T_2^*; t)\, dt. \tag{8.32}$$

Note that when $\lambda(\infty; x) = 1$, a finite T_2^* always exists.

8.3 Cumulative Damage Model with Random Failure Level

A unit is subjected to shocks and suffers some damage due to shocks. Each damage is additive and the unit fails when the total damage has exceeded a failure level K $(0 < K < \infty)$. It is assumed that shocks occur at a renewal process $\{X_j\}$ with an identical interarrival distribution $\Pr\{X_j \leq t\} = F(t)$ with finite mean $1/\lambda$ $(0 < \lambda < \infty)$. An amount of damage W_j due to the jth $(j = 1, 2, \ldots)$ shock is independent of X_j and has an identical distribution $W(x) \equiv \Pr\{W_j \leq x\}$ with finite mean $1/\omega$ $(0 < \omega < \infty)$. In general, $\Phi^{(j)}(t)$ denotes the j-fold Stieltjes convolution of $\Phi(t)$ with itself, and $\Phi^{(0)}(t) \equiv 1$ for $t \geq 0$.

8.3.1 Replacement Policy

Suppose that the unit is replaced at a planned time T $(0 < T \leq \infty)$, at a shock number N $(N = 1, 2, \ldots)$, or at a failure level K, whichever occurs first. Let c_T, c_N and c_K be the respective replacement costs for time T, shock N, and level K with $c_K > c_T$ and $c_K > c_N$. Then, the expected cost rate is, from [1, p. 42],

$$C_1(T, N; K) = \frac{c_K - (c_K - c_N)F^{(N)}(T)W^{(N)}(K) - (c_K - c_T)\sum_{j=0}^{N-1}[F^{(j)}(T) - F^{(j+1)}(T)]W^{(j)}(K)}{\sum_{j=0}^{N-1} \int_0^T [F^{(j)}(t) - F^{(j+1)}(t)]\, dt\, W^{(j)}(K)}. \tag{8.33}$$

In addition, when a failure level K is not constant and has a general distribution $G(x) \equiv \Pr\{K \le x\}$ with finite mean $1/\theta$ $(0 < \theta < \infty)$ [10, p. 164],

$$C_1(T, N; G) = \frac{\begin{aligned} c_K &- (c_K - c_N) F^{(N)}(T) \int_0^\infty W^{(N)}(x) \, dG(x) \\ &- (c_K - c_T) \sum_{j=0}^{N-1} [F^{(j)}(T) - F^{(j+1)}(T)] \int_0^\infty W^{(j)}(x) \, dG(x) \end{aligned}}{\sum_{j=0}^{N-1} \int_0^T [F^{(j)}(t) - F^{(j+1)}(t)] \, dt \int_0^\infty W^{(j)}(x) \, dG(x)}.$$

$$(8.34)$$

8.3.1.1 Optimum Time

Suppose that the unit is replaced only at time T or at failure, whichever occurs first. Then, the expected cost rate is, from (8.34),

$$
\begin{aligned}
C_1(T; G) &\equiv \lim_{N \to \infty} C_1(T, N; G) \\
&= \frac{c_K - (c_K - c_T) \sum_{j=0}^{\infty} [F^{(j)}(T) - F^{(j+1)}(T)] \int_0^\infty W^{(j)}(x) \, dG(x)}{\sum_{j=0}^{\infty} \int_0^T [F^{(j)}(t) - F^{(j+1)}(t)] \, dt \int_0^\infty W^{(j)}(x) \, dG(x)}.
\end{aligned}
$$

$$(8.35)$$

When $G(x) = 1 - e^{-\theta x}$,

$$C_1(T; \theta) = \frac{c_K - (c_K - c_T) \sum_{j=0}^{\infty} [F^{(j)}(T) - F^{(j+1)}(T)][W^*(\theta)]^j}{\sum_{j=0}^{\infty} \int_0^T [F^{(j)}(t) - F^{(j+1)}(t)] \, dt [W^*(\theta)]^j}. \qquad (8.36)$$

Let $f(t)$ be a density function $F(t)$, $f^{(j)}(t)$ $(j = 1, 2, \ldots)$ be the j-fold Stieltjes convolution of $f(t)$ with itself, and $f^{(0)}(t) \equiv 0$ for $t \ge 0$, i.e., $f^{(j)}(t) \equiv dF^{(j)}(t)/dt$. Then, differentiating $C_1(T; \theta)$ with respect to T and setting it equal to zero,

$$Q_1(T) \sum_{j=0}^{\infty} [W^*(\theta)]^j \int_0^T [F^{(j)}(t) - F^{(j+1)}(t)] \, dt$$

$$+ \sum_{j=0}^{\infty} [W^*(\theta)]^j [F^{(j)}(T) - F^{(j+1)}(T)] = \frac{c_K}{c_K - c_T}, \qquad (8.37)$$

where

$$Q_1(T) \equiv \frac{\sum_{j=0}^{\infty} [W^*(\theta)]^j [f^{(j+1)}(T) - f^{(j)}(T)]}{\sum_{j=0}^{\infty} [W^*(\theta)]^j [F^{(j)}(T) - F^{(j+1)}(T)]}.$$

If $Q_1(T)$ increases strictly with T, then the left-hand side of (8.37) also increases strictly with T from 1 to $Q_1(\infty)/\{\lambda[1-W^*(\theta)]\}$, where $Q_1(\infty) \equiv \lim_{T\to\infty} Q_1(T)$. Thus, if $Q_1(\infty) > \lambda[1-W^*(\theta)][c_K/(c_K-c_T)]$, then there exists a finite and unique T^* $(0 < T^* < \infty)$ which satisfies (8.37).

Furthermore, when shocks occur at a nonhomogeneous Poisson process with cumulative hazard rate $H(t) \equiv \int_0^t h(u)\,du$, i.e., $F^{(j)}(t) = \sum_{i=j}^{\infty}\{[H(t)]^i/i!\}e^{-H(t)}$ $(j = 0, 1, 2, \ldots)$, the expected cost rate in (8.36) is [1, p. 43]

$$C_1(T;\theta) = \frac{c_K - (c_K - c_T)e^{-[1-W^*(\theta)]H(T)}}{\int_0^T e^{-[1-W^*(\theta)]H(t)}\,dt}. \tag{8.38}$$

In this case, (8.37) becomes

$$[1 - W^*(\theta)]h(T)\int_0^T e^{-[1-W^*(\theta)]H(t)}\,dt + e^{-[1-W^*(\theta)]H(T)} = \frac{c_K}{c_K - c_T}. \tag{8.39}$$

This corresponds to an age replacement policy with a failure distribution $(1 - \exp\{-[1 - W^*(\theta)]H(t)\})$ in (2.2). Thus, if $h(t)$ increases strictly to $h(\infty) \equiv \lim_{t\to\infty} h(t)$, then the left-hand side of (8.39) also increases strictly from 1 to

$$L_2(\infty) \equiv [1 - W^*(\theta)]h(\infty)\int_0^{\infty} e^{-[1-W^*(\theta)]H(t)}\,dt.$$

Therefore, we have the following optimum policy:

(i) If $L_2(\infty) > c_K/(c_K - c_T)$, then there exists a finite and unique T^* $(0 < T^* < \infty)$ which satisfies (8.39), and the resulting cost rate is

$$C_1(T^*;\theta) = (c_K - c_T)[1 - W^*(\theta)]h(T^*). \tag{8.40}$$

(ii) If $L_2(\infty) \leq c_K/(c_K - c_T)$, then $T^* = \infty$, and the resulting cost rate is

$$C_1(\infty;\theta) \equiv \lim_{T\to\infty} C_1(T;\theta) = \frac{c_K}{\int_0^{\infty} e^{-[1-W^*(\theta)]H(t)}\,dt}. \tag{8.41}$$

8.3.1.2 Optimum Number

Suppose that the unit is replaced only at shock N $(N = 1, 2, \ldots)$ or at failure, whichever occurs first. Then, the expected cost rate is, from (8.34),

$$C_1(N; \theta) \equiv \lim_{T \to \infty} C_1(T, N; \theta)$$

$$= \frac{c_K - (c_K - c_N) \int_0^\infty W^{(N)}(x) \, dG(x)}{(1/\lambda) \sum_{j=0}^{N-1} \int_0^\infty W^{(j)}(x) \, dG(x)} \qquad (N = 1, 2, \ldots). \qquad (8.42)$$

In particular, when $N = 1$, i.e., the unit is always replaced at the first shock, the expected cost rate is

$$C_1(1; \theta) = \lambda \left[c_K - (c_K - c_N) \int_0^\infty W(x) \, dG(x) \right]. \qquad (8.43)$$

Forming the inequality $C_1(N + 1; \theta) - C_1(N; \theta) \geq 0$,

$$Q_2(N) \sum_{j=0}^{N-1} \int_0^\infty W^{(j)}(x) \, dG(x) + \int_0^\infty W^{(N)}(x) \, dG(x) \geq \frac{c_K}{c_K - c_N}, \qquad (8.44)$$

where

$$Q_2(N) \equiv \frac{\int_0^\infty [W^{(N)}(x) - W^{(N+1)}(x)] \, dG(x)}{\int_0^\infty W^{(N)}(x) \, dG(x)}.$$

If $Q_2(N)$ increases strictly, i.e., $\int_0^\infty W^{(N+1)}(x) \, dG(x) / \int_0^\infty W^{(N)}(x) \, dG(x)$ decreases strictly with N, then the left-hand side of (8.44) increases strictly to

$$Q_2(\infty) \int_0^\infty [1 + M(x)] \, dG(x),$$

where $M(x) \equiv \sum_{j=1}^\infty W^{(j)}(x)$ and $Q_2(\infty) \equiv \lim_{N \to \infty} Q_2(N) \leq 1$. Thus, if $Q_2(\infty) \int_0^\infty [1 + M(x)] \, dG(x) > c_K / (c_K - c_N)$, then there exists a finite and unique minimum N^* ($1 \leq N^* < \infty$) which satisfies (8.44), and the resulting cost rate is

$$\lambda(c_K - c_N) Q_2(N^*) < C_2(N^*) \leq \lambda(c_K - c_N) Q_2(N^* + 1). \qquad (8.45)$$

Conversely, if $Q_2(\infty) \int_0^\infty [1 + M(x)] \, dG(x) \leq c_K / (c_K - c_N)$, then $N^* = \infty$. Because $Q_2(N)$ represents the probability that the unit surviving at shock N will fail at shock $N+1$, $Q_2(N+1)$ would increase to 1. In this case, if $\int_0^\infty [1 + M(x)] \, dG(x) > c_K / (c_K - c_N)$, i.e., the expected number of shocks until failures is greater than $c_K / (c_K - c_N)$, then a finite N^* exists uniquely. When $G(x) = 1 - e^{-\theta x}$, $Q_2(N) = 1 - W^*(\theta)$, and hence, $N^* = \infty$.

In particular, when $W(x) = 1 - e^{-\omega x}$,

$$Q_2(N) = \frac{\int_0^\infty [(\omega x)^N/N!] e^{-\omega x} \, dG(x)}{\sum_{j=N}^\infty \int_0^\infty [(\omega x)^j/j!] e^{-\omega x} \, dG(x)}$$

$$= 1 \Bigg/ \left[1 + \frac{\sum_{j=N+1}^\infty \int_0^\infty [(\omega x)^j/j!] e^{-\omega x} \, dG(x)}{\int_0^\infty [(\omega x)^N/N!] e^{-\omega x} \, dG(x)} \right]$$

$$= 1 \Bigg/ \left[1 + \frac{\omega \int_0^\infty [(\omega x)^N/N!] e^{-\omega x} \overline{G}(x) \, dx}{\int_0^\infty [(\omega x)^N/N!] e^{-\omega x} \, dG(x)} \right].$$

It can be proved from (3) of Appendix A.1 that when $r(x) \equiv g(x)/\overline{G}(x)$ increases strictly to $r(\infty)$, where $g(x)$ is a density function of $G(x)$, $Q_2(N)$ increases strictly to $r(\infty)/[\omega + r(\infty)]$. Thus, if $\{r(\infty)/[\omega + r(\infty)]\}(1 + \omega/\theta) > c_K/(c_K - c_N)$ then a finite N^* ($1 \le N^* < \infty$) exists. In addition, when $r(\infty) = \infty$, i.e., $Q_2(\infty) = 1$, if $\omega/\theta > c_N/(c_K - c_N)$ then a finite N^* exists.

8.3.2 Periodic Replacement

It is assumed that the total damage due to shocks is additive when it has not exceeded a failure level K, however, it is not additive at any shock after it has exceeded K. In this case, minimal maintenance is done at each shock and the damage level remains in K. Suppose that the unit is replaced at a planned time T ($0 < T \le \infty$) or at a shock number N ($N = 1, 2, \ldots$), whichever occurs first. Let c_M be the cost of minimal maintenance. Then, the expected number of minimal maintenance, i.e., the expected number of shocks in the case where the total damage remains in K after it has reached K, is [1, p. 151] (Problem 9 in Sect. 8.5)

$$\sum_{j=0}^{N-1} \Bigg([W^{(j)}(K) - W^{(j+1)}(K)] \Bigg\{ \sum_{i=0}^{N-1-j} i[F^{(i+j)}(T) - F^{(i+j+1)}(T)]$$

$$+ (N - 1 - j)F^{(N)}(T) \Bigg\} \Bigg) = \sum_{j=0}^{N-1} [1 - W^{(j)}(K)]F^{(j)}(T).$$

Thus, by the similar method of obtaining (8.33), the expected cost rate is

$$C_2(T, N; K) = \frac{c_T + (c_N - c_T)F^{(N)}(T) + c_M \sum_{j=0}^{N-1}[1 - W^{(j)}(K)]F^{(j)}(T)}{\sum_{j=0}^{N-1} \int_0^T [F^{(j)}(t) - F^{(j+1)}(t)] \, dt},$$

$$(8.46)$$

which agrees with (4.22) of [8, p. 104] when $K = 0$ and $F^{(j)}(t) = \sum_{i=j}^\infty p_i(t)$.

In addition, when a failure level K is not constant and has a general distribution $G(x) \equiv \Pr\{K \le x\}$,

$$C_2(T, N; G) = \frac{\begin{aligned}&c_T + (c_N - c_T)F^{(N)}(T) \\ &+ c_M \sum_{j=0}^{N-1} F^{(j)}(T) \int_0^T [1 - W^{(j)}(x)]\, dG(x)\end{aligned}}{\sum_{j=0}^{N-1} \int_0^T [F^{(j)}(t) - F^{(j+1)}(t)]}. \tag{8.47}$$

8.3.2.1 Optimum Time

Suppose that the unit is replaced only at time T, i.e., the unit is replaced at periodic times kT ($k = 1, 2, \ldots$). Then, the expected cost rate is, from (8.47),

$$\begin{aligned} C_2(T; G) &\equiv \lim_{N \to \infty} C_2(T, N; G) \\ &= \frac{c_T + c_M \sum_{j=1}^{\infty} F^{(j)}(T) \int_0^{\infty} [1 - W^{(j)}(x)]\, dG(x)}{T}. \end{aligned} \tag{8.48}$$

We find an optimum T^* which minimizes $C_2(T; G)$. Differentiating $C_2(T; G)$ with respect to T and setting it equal to zero,

$$\sum_{j=1}^{\infty} \int_0^T t\, df^{(j)}(t) \int_0^{\infty} [1 - W^{(j)}(x)]\, dG(x) = \frac{c_T}{c_M}. \tag{8.49}$$

Thus, if

$$\sum_{j=1}^{\infty} [f^{(j)}(T)]' \int_0^{\infty} [1 - W^{(j)}(x)]\, dG(x) > 0,$$

and

$$\sum_{j=1}^{\infty} \int_0^{\infty} t\, df^{(j)}(t) \int_0^{\infty} [1 - W^{(j)}(x)]\, dG(x) > \frac{c_T}{c_M},$$

then there exists a finite and unique T^* ($0 < T^* < \infty$) which satisfies (8.49).

In particular, when $F(t) = 1 - e^{-\lambda t}$, (8.49) is (Problem 10 in Sect. 8.5)

$$\sum_{j=1}^{\infty} \frac{(\lambda T)^j}{j!} e^{-\lambda T} \sum_{i=1}^{j} \int_0^{\infty} [W^{(i)}(x) - W^{(j)}(x)]\, dG(x) = \frac{c_T}{c_M}, \tag{8.50}$$

whose left-hand side increases strictly from 0 to $\int_0^{\infty} M(x)\, dG(x)$. Therefore, if $\int_0^{\infty} M(x)\, dG(x) > c_T/c_M$, then there exists a finite and unique T^* ($0 < T^* < \infty$) which satisfies (8.50). In addition, when $G(t) = 1 - e^{-\theta t}$, (8.50) is

$$\frac{W^*(\theta)}{1 - W^*(\theta)} \left(1 - \{1 + [1 - W^*(\theta)]\lambda T\}e^{-[1-W^*(\theta)]\lambda T}\right) = \frac{c_T}{c_M}, \tag{8.51}$$

whose left-hand increases strictly from 0 to $M^*(\theta) = W^*(\theta)/[1 - W^*(\theta)]$.

8.3.2.2 Optimum Number

Suppose that the unit is replaced only at shock N ($N = 1, 2, \ldots$). Then, the expected cost rate is, from (8.47),

$$
\begin{aligned}
C_2(N; G) &\equiv \lim_{T \to \infty} C_2(T, N; G) \\
&= \frac{c_N + c_M \sum_{j=0}^{N-1} \int_0^\infty [1 - W^{(j)}(x)] \, dG(x)}{N/\lambda} \quad (N = 1, 2, \ldots). \tag{8.52}
\end{aligned}
$$

Forming the inequality $C_2(N + 1; G) - C_2(N; G) \geq 0$,

$$\sum_{j=0}^{N-1} \int_0^\infty [W^{(j)}(x) - W^{(N)}(x)] \, dG(x) \geq \frac{c_N}{c_M}, \tag{8.53}$$

whose left-hand side increases strictly with N to $\int_0^\infty [1 + M(x)] \, dG(x)$. Thus, if $\int_0^\infty [1 + M(x)] \, dG(x) > c_N/c_M$, then there exists a finite and unique minimum N^* ($1 \leq N^* < \infty$) which satisfies (8.53). In particular, when $G(x) = 1 - e^{-\theta x}$, (8.53) is

$$\sum_{j=0}^{N-1} \{[W^*(\theta)]^j - [W^*(\theta)]^N\} \geq \frac{c_N}{c_M}, \tag{8.54}$$

which increases strictly with N from $1 - W^*(\theta)$ to $1/[1 - W^*(\theta)]$. If $1 - W^*(\theta) \geq c_N/c_M$, then $N^* = 1$, and if $1 - W^*(\theta) \geq c_M/c_N$ then $N^* = \infty$. For example, when $c_N = c_M$, a finite N^* ($2 \leq N^* < \infty$) always exists.

8.3.3 Continuous Damage Model

The continuous damage $Z(t)$ usually increases swaying with time from $Z(0) = 0$, and the unit fails when $Z(t)$ has exceeded a failure level K [10, p. 184]: First, it is assumed that $Z(t) = A(t)t$, where $A(t)$ is normally distributed with mean α ($\alpha > 0$) and variance σ^2/t. Then, the reliability at time t is

$$R(t) = \Pr\{A(t) \leq K/t\} = \Phi\left(\frac{K - \alpha t}{\sigma \sqrt{t}}\right), \tag{8.55}$$

where $\Phi(x)$ is the standard normal distribution with mean 0 and variance 1, i.e.,
$\Phi(x) \equiv (1/\sqrt{2\pi}) \int_{-\infty}^{x} e^{-u^2/2} \, du$.

Suppose that the unit is replaced at time T $(0 < T \le \infty)$ or at failure, whichever occurs first. Then, the expected cost rate is, from (2.2),

$$C_3(T, K) = \frac{c_K - (c_K - c_T)\Phi[(K - \alpha T)/\sigma\sqrt{T}]}{\int_0^T \Phi[(K - \alpha t)/\sigma\sqrt{t}] \, dt}, \tag{8.56}$$

where c_T and c_K are given in (8.33). Furthermore, when a failure level K is a random variable with a general distribution $G(x) \equiv \Pr\{K \le x\}$, the expected cost rate in (8.56) is

$$C_3(T; G) = \frac{c_K - (c_K - c_T) \int_0^\infty \Phi[(x - \alpha T)/\sigma\sqrt{T}] \, dG(x)}{\int_0^T \{\int_0^\infty \Phi[(x - \alpha t)/\sigma\sqrt{t}] \, dG(x)\} \, dt}. \tag{8.57}$$

When $\alpha = 1$ and $\sigma = 1$, differentiating $C_3(T; G)$ with T and setting it equal to zero,

$$\frac{\int_0^\infty (x/T + 1)\phi(x/\sqrt{T} - \sqrt{T}) \, dG(x)}{2\sqrt{T} \int_0^\infty \Phi(x/\sqrt{T} - \sqrt{T}) \, dG(x)} \int_0^T \left[\int_0^\infty \Phi\left(\frac{x}{\sqrt{t}} - \sqrt{t}\right) dG(x) \right] dt$$

$$+ \int_0^\infty \Phi\left(\frac{x}{\sqrt{T}} - \sqrt{T}\right) dG(x) = \frac{c_K}{c_K - c_T}, \tag{8.58}$$

where $\phi(x) \equiv d\Phi(x)/dx = (1/\sqrt{2\pi})e^{-x^2/2}$ (Problem 11 in Sect. 8.5).

Next, it is assumed that $Z(t) = \alpha t + B(t)$, where $B(t)$ has an exponential distribution $1 - e^{-x/\sigma\sqrt{t}}$ [10, p. 192]. That is, the total damage increases linearly with time t, however, it undergoes positively some damage according to an exponential distribution with mean $\sigma\sqrt{t}$. Then, the reliability at time t is

$$R(t) = \Pr\{B(t) \le K - \alpha t\} = 1 - \exp\left(-\frac{K - \alpha t}{\sigma\sqrt{t}}\right). \tag{8.59}$$

When the unit is replaced before failure at time T, the expected cost rate is

$$C_4(T; K) = \frac{c_K - (c_K - c_T)\{1 - \exp[-(K - \alpha T)/\sigma\sqrt{T}]\}}{\int_0^T \{1 - \exp[-(K - \alpha t)/\sigma\sqrt{t}]\} \, dt}. \tag{8.60}$$

Furthermore, when K has a general distribution $G(x)$, (8.60) becomes

$$C_4(T; G) = \frac{c_K - (c_K - c_T) \int_0^\infty \{1 - \exp[-(x - \alpha T)/\sigma\sqrt{T}]\} \, dG(x)}{\int_0^T (\int_0^\infty \{1 - \exp[-(x - \alpha t)/\sigma\sqrt{t}]\} \, dG(x)) \, dt}. \tag{8.61}$$

Differentiating $C_4(T; G)$ with respect to T and setting it equal to zero (Problem 12 in Sect. 8.5),

$$\frac{\int_0^\infty (x/T + \alpha) \exp[-(x - \alpha T)/\sigma \sqrt{T}] \, dG(x)}{2\sigma \sqrt{T} \int_0^\infty \{1 - \exp[-(x - \alpha T)/\sigma \sqrt{T}]\} \, dG(x)}$$

$$\times \int_0^T \left\{ \int_0^\infty \left[1 - \exp\left(-\frac{x - \alpha t}{\sigma \sqrt{t}} \right) \right] dG(x) \right\} dt$$

$$+ \int_0^\infty \left[1 - \exp\left(-\frac{x - \alpha T}{\sigma \sqrt{T}} \right) \right] dG(x) = \frac{c_K}{c_K - c_T}. \qquad (8.62)$$

8.4 Other Random Reliability Models

For further studies, we propose briefly the following three random models:

8.4.1 Random Scheduling Time

It is assumed in N tandem works of Sect. 7.1.2 that L is not constant and is a random variable with a general distribution $\Pr\{L \le t\} \equiv A(t)$ with finite mean $1/l$ ($0 < l < \infty$). Then, the total expected cost in (7.12) is rewritten as

$$C(N; A) = c_S \int_0^\infty [1 - G^{(N)}(t)] A(t) \, dt$$

$$+ c_E \int_0^\infty G^{(N)}(t) \overline{A}(t) \, dt + \frac{c_0}{l} \quad (N = 0, 1, 2, \ldots). \qquad (8.63)$$

We find an optimum N^* which minimizes $C(N; A)$. From the inequality $C(N + 1; A) - C(N; A) \ge 0$,

$$\frac{\int_0^\infty [G^{(N)}(t) - G^{(N+1)}(t)] A(t) \, dt}{\int_0^\infty [G^{(N)}(t) - G^{(N+1)}(t)] \overline{A}(t) \, dt} \ge \frac{c_E}{c_S},$$

i.e.,

$$\theta \int_0^\infty [G^{(N)}(t) - G^{(N+1)}(t)] \overline{A}(t) \, dt \le \frac{c_S}{c_S + c_E}. \qquad (8.64)$$

In particular, when $A(t) = 1 - e^{-lt}$, (8.64) becomes

$$\frac{\theta[1 - G^*(l)][G^*(l)]^N}{l} \le \frac{c_S}{c_S + c_E},$$
(8.65)

whose left-hand decreases strictly with N from $\theta[1 - G^*(l)]/l$ to 0. Thus, there exists a finite and unique minimum N^* ($0 \le N^* < \infty$) which satisfies (8.65). If $\theta[1 - G^*(l)]/l \le c_S/(c_S + c_E)$, then $N^* = 0$. In addition, when $G(t) = 1 - e^{-\theta t}$, (8.65) is simplified as

$$\left(\frac{\theta}{\theta + l}\right)^{N+1} \le \frac{c_S}{c_S + c_E},$$

whose left-hand decreases with l from 1 to 0, i.e., N^* increases with $1/l$ from 0 to ∞, and decreases with $1/\theta$ from ∞ to 0.

8.4.2 Random Inspection Number

It has been proposed in inspection policies [3, p. 181] that the unit is checked at periodic times kT ($k = 1, 2, \ldots, N-1$) and is replaced at time NT ($N = 1, 2, \ldots$), as an example of missiles [8, p. 204]. When the unit has a failure distribution $F(t)$, the expected cost rate is [3, p. 182]

$$C(T; N) = \frac{c_I \sum_{k=0}^{N-1} \overline{F}(kT) - c_D \int_0^{NT} \overline{F}(t)\,dt + c_R}{T \sum_{k=0}^{N-1} \overline{F}(kT)} + c_D,$$
(8.66)

where c_I = cost for one check, c_D = cost per unit of time for the time elapsed between a failure and its detection at the next check time, and c_R = replacement cost at time NT or at failure.

It is assumed that N is a random variable with a probability function $p_k \equiv \Pr\{N = k\}$ ($k = 1, 2, \ldots$) and $\sum_{k=1}^{\infty} p_k = 1$. Then, the expected cost rate in (8.66) is

$$C(T; p) = \frac{c_I \sum_{k=0}^{\infty} \overline{P}_k \overline{F}(kT) + c_D \sum_{k=0}^{\infty} \overline{P}_k \int_{kT}^{(k+1)T} [\overline{F}(kT) - \overline{F}(t)]\,dt + c_R}{T \sum_{k=0}^{\infty} \overline{P}_k \overline{F}(kT)},$$
(8.67)

where $\overline{P}_k \equiv \sum_{j=k+1}^{\infty} p_j$ ($k = 0, 1, 2, \ldots$) and $\overline{P}_0 = 1$.

In particular, when $F(t) = 1 - e^{-\lambda t}$ ($0 < \lambda < \infty$) and $p_k = pq^{k-1}$, i.e., $\overline{P}_k = q^k$, where $0 < p \le 1$ and $q \equiv 1 - p$,

$$C(T; p) = \frac{1}{T}\left\{c_I + c_D\left[T - \frac{1}{\lambda}(1 - e^{-\lambda T})\right] + c_R(1 - qe^{-\lambda T})\right\}.$$
(8.68)

Differentiating $C(T; p)$ with respect to T and setting it equal to zero,

$$\frac{c_D}{\lambda}[1 - (1 + \lambda T)e^{-\lambda T}] - c_R[1 - q(1 + \lambda T)e^{-\lambda T}] = c_I, \qquad (8.69)$$

whose left-hand side increases strictly with T from $-c_R p$ to $c_D/\lambda - c_R$ for $c_R q < c_D/\lambda$. Thus, there exists a finite and unique T_q^* $(0 < T_q^* < \infty)$ which satisfies (8.69) for $c_D/\lambda > c_R + c_I$. In this case, T_q^* decreases with q from T_0^* which is a solution of the equation

$$1 - (1 + \lambda T)e^{-\lambda T} = \frac{c_I + c_R}{c_D/\lambda}. \qquad (8.70)$$

Note that (8.70) agrees with (8.103) of standard periodic inspection [3, p. 183].

8.4.3 Random Number of Failures

The unit begins to operate at time 0 and undergoes only minimal repair at failures. Suppose that the unit is replaced at time T or at failure N $(N = 1, 2, \ldots)$, whichever occurs first. Then, because the probability that j failures occur exactly in $[0, t]$ is $p_j(t) \equiv \{[H(t)^j/j!]\}e^{-H(t)}$ $(j = 0, 1, 2, \ldots)$ [8, p. 97], the expected cost rate is [3, p. 163]

$$C_F(T; N) = \frac{c_M[N - 1 - \sum_{j=0}^{N-1}(N - 1 - j)p_j(T)] + c_R}{\sum_{j=0}^{N-1}\int_0^T p_j(t)\, dt}, \qquad (8.71)$$

where c_M = cost for minimal repair and c_R = replacement cost at time T or at failure N.

Next, N is a random variable with a probability function p_k denoted in Sect. 8.4.2. Then, the expected cost rate in (8.71) is

$$C_F(T; p) = \frac{c_M \sum_{k=1}^{\infty} p_k(T)[k - \sum_{j=1}^{k}(k - j + 1)p_j] + c_R}{\sum_{k=0}^{\infty} \overline{P}_k \int_0^T p_k(t)\, dt}. \qquad (8.72)$$

In particular, when $p_k = pq^{k-1}$ $(k = 1, 2, \ldots)$ $(0 < p < 1)$,

$$\begin{aligned}
C_F(T; p) &= \frac{c_M \sum_{k=1}^{\infty} p_k(T) \sum_{j=1}^{k} q^j + c_R}{\sum_{k=0}^{\infty} q^k \int_0^T p_k(t)\, dt} \\
&= \frac{c_M(q/p)[1 - e^{-pH(T)}] + c_R}{\int_0^T e^{-pH(t)}\, dt}. \qquad (8.73)
\end{aligned}$$

It is of interest that $C_F(T; p)$ in (8.73) corresponds to an age replacement with a failure distribution $F_p(t) = 1 - e^{-pH(t)}$, where the replacement cost when the unit is replaced at failure N is $c_M(q/p)$, and the replacement cost at time T is c_R. Therefore, if $h(t)$ increases strictly to $h(\infty) = \infty$, then there exists a finite and unique T^* $(0 < T^* < \infty)$ which satisfies

$$h(T) \int_0^T e^{-pH(t)}\, dt - \frac{1 - e^{-pH(T)}}{p} = \frac{c_R}{c_M q}. \tag{8.74}$$

Finally, suppose that the unit is replaced at time T or at failure N, whichever occurs last. Then, the mean time to replacement is

$$T \sum_{j=N}^{\infty} p_j(T) + \int_T^{\infty} t\, d\left[\sum_{j=N}^{\infty} p_j(t) \right] = T + \sum_{j=0}^{N-1} \int_T^{\infty} p_j(t)\, dt,$$

and the expected number of minimal repairs is

$$\sum_{j=N}^{\infty} j\, p_j(T) + (N-1) \sum_{j=0}^{N-1} p_j(T) = N - 1 + \sum_{j=N-1}^{\infty} (j - N + 1)\, p_j(T).$$

Thus, the expected cost rate is

$$C_L(T; N) = \frac{c_M\left[N - 1 + \sum_{j=N-1}^{\infty} (j - N + 1)\, p_j(T) \right] + c_R}{T + \sum_{j=0}^{N-1} \int_T^{\infty} p_j(t)\, dt}. \tag{8.75}$$

Furthermore, when N is a random variable with a probability function p_k, the expected cost rate is (Problem 13 in Sect. 8.5)

$$C_L(T; p) = \frac{c_M \sum_{k=0}^{\infty} p_k(T)\left[k + \sum_{j=k+1}^{\infty} (j - k - 1)\, p_j \right] + c_R}{T + \sum_{k=0}^{\infty} \overline{P}_{k+1} \int_T^{\infty} p_k(t)\, dt}, \tag{8.76}$$

(Problems 14 and 15 in Sect. 8.5).

8.5 Problems

1. Show that

$$
\sum_{k=0}^{\infty} \overline{G}[(k+1)T] \int_{kT}^{(k+1)T} dF(t) + \sum_{k=0}^{\infty} \int_{kT}^{(k+1)T} \left[\int_{t}^{(k+1)T} dG(u) \right] dF(t)
$$

$$
+ \sum_{k=0}^{\infty} \int_{kT}^{(k+1)T} \overline{F}(t) \, dG(t) = 1.
$$

2. Derive (8.1).
3. Prove that the left-hand of (8.3) decreases strictly with θ from $[e^{\lambda T} - (1 + \lambda T)]/\lambda$ to 0.
4. Explain why T^* becomes about 2 times when c_T becomes 4 times, independently of $1/\theta$.
5. Derive (8.7) and (8.8).
6. Show that T^* decreases with $1/\theta$ to T_S^*, and compute T^* numerically when $F(t) = 1 - \exp[-(\lambda t)^{\alpha}]$ $(\alpha > 1)$.
7. Compute T_k^* when $B_{k+1} \equiv \prod_{j=1}^{k}[1 + j/(j+1)]$ $(k = 1, 2, \ldots)$ and $B_1 \equiv 1$.
8. Compute T^* which satisfies (8.18) and T_k^* which satisfy (8.21).
9. Show that the expected number of shocks is

$$
\sum_{j=0}^{N-1}[1 - W^{(j)}(K)]F^{(j)}(T).
$$

10. Derive (8.50) and (8.51).
11. Compute numerically optimum T^* which satisfies (8.58).
12. Compute numerically optimum T^* which satisfies (8.62).
* 13. Discuss optimum policies which minimize $C_F(T; p)$ in (8.72) and $C_L(T; p)$ in (8.76).
* 14. Consider the block replacement where the unit is replaced at time T and failure N, whichever occurs first or last, and N is a random variable [3, p. 173].
* 15. Make other random models by transforming constant number to random one.

References

1. Nakagawa T (2007) Shock and damage models in reliability theory. Springer, London
2. Yun WY, Choi CH (2000) Optimum replacement intervals with random time horizon. J Qual Mainte Eng 6:269–274
3. Nakagawa T (2008) Advanced reliability models and maintenance policies. Springer, London

4. Nakagawa T, Yasui K, Sandoh H (2004) Note on optimal partition problems in reliability models. J Qual Mainte Eng 10:282–287
5. Nakagawa T, Mizutani S (2009) A summary of maintenance policies for a finite interval. Reliab Eng Syst Saf 94:89–96
6. Yun WY, Nakagawa T (2010) Replacement and inspection policies for products with random life cycle. Reliab Eng Syst Saf 95:161–165
7. Barlow RE, Proschan F (1965) Mathematical theory of reliability. Wiley, New York
8. Nakagawa T (2005) Maintenance theory of reliability. Springer, London
9. Wang H, Pham H (2003) Optimum imperfect maintenance models. In: Pham H (ed) Handbook of reliability engineering. Springer, London, pp 397–414
10. Nakagawa T (2011) Stochastic processes with applications to reliability theory. Springer, London

Appendix A
Extended Failure Rates

Suppose that the unit operates for a job with random working times. It is assumed that the unit has a failure distribution $F(t)$ for $t \geq 0$ with finite mean $\mu \equiv \int_0^\infty \overline{F}(t)dt < \infty$, where $\overline{\Phi}(t) \equiv 1 - \Phi(t)$ for any function $\Phi(t)$. When $F(t)$ has a density function $f(t) \equiv dF(t)/dt$, i.e., $F(t) \equiv \int_0^t f(u)du$, the failure rate $h(t) \equiv f(t)/\overline{F}(t)$ for $F(t) < 1$ is assumed to increase from $h(0) \equiv \lim_{t \to 0} h(t)$ to $h(\infty) \equiv \lim_{t \to \infty} h(t)$. In addition, the working time of a job has an exponential distribution $(1 - e^{-\theta t})$ with finite mean $1/\theta < \infty$.

A.1 Properties of Failure Rates $f(t)/\overline{F}(t)$

This appendix investigates the properties of extended failure rates appeared in this book and summarizes them.

(1) For $0 < T < \infty$ and $N = 0, 1, 2, \ldots$,

$$r_{N+1}(T) = \frac{\theta(\theta T)^N/N!}{\sum_{j=0}^{N}[(\theta T)^j/j!]}$$

increases strictly with T from 0 to θ for $N \geq 1$, is θ for $N = 0$, and decreases strictly with N from θ to 0.

Proof For $N \geq 1$,

$$\lim_{T \to 0} r_{N+1}(T) = 0, \quad \lim_{T \to \infty} r_{N+1}(T) = \theta.$$

© Springer-Verlag London 2014
T. Nakagawa, *Random Maintenance Policies*,
Springer Series in Reliability Engineering, DOI 10.1007/978-1-4471-6575-0

Differentiating $r_{N+1}(T)$ with respect to T,

$$\frac{dr_{N+1}(T)}{dT} = \frac{\theta^2}{\{\sum_{j=0}^{N}[(\theta T)^j/j!]\}^2} \frac{(\theta T)^{N-1}}{N!} \sum_{j=0}^{N} \frac{(\theta T)^j}{j!}(N-j) > 0,$$

which implies that $r_{N+1}(T)$ increases strictly with T from 0 to θ. When $N = 0$, $r_1(T) = \theta$ which is constant for any $T > 0$. Similarly, for $0 < T < \infty$,

$$\lim_{N \to 0} r_{N+1}(T) = \theta, \qquad \lim_{N \to \infty} r_{N+1}(T) = 0,$$

$$r_{N+1}(T) - r_N(T) = \frac{\theta}{\sum_{j=0}^{N}[(\theta T)^j/j!] \sum_{j=0}^{N-1}[(\theta T)^j/j!]}$$

$$\times \frac{(\theta T)^{N-1}}{N!} \sum_{j=0}^{N} \frac{(\theta T)^j}{j!}(j-N) < 0,$$

which implies that $r_{N+1}(T)$ decreases strictly with N from θ to 0.

(2) For $0 < T < \infty$ and $N = 0, 1, 2, \ldots$,

$$\widetilde{r}_{N+1}(T) = \frac{\theta(\theta T)^N/N!}{\sum_{j=N}^{\infty}[(\theta T)^j/j!]}$$

decreases strictly with T from θ to 0 for $N \geq 1$ and is $\theta e^{-\theta T}$ for $N = 0$, and increases strictly with N from $\theta e^{-\theta T}$ to θ.

Proof For $N \geq 1$,

$$\lim_{T \to 0} \widetilde{r}_{N+1}(T) = \theta, \quad \lim_{T \to \infty} \widetilde{r}_{N+1}(T) = 0.$$

Differentiating $\widetilde{r}_{N+1}(T)$ with respect to T,

$$\frac{d\widetilde{r}_{N+1}(T)}{dT} = \frac{\theta^2}{\{\sum_{j=N}^{\infty}(\theta T)^j/j!\}^2} \frac{(\theta T)^{N-1}}{N!} \sum_{j=N}^{\infty} \frac{(\theta T)^j}{j!}(N-j) < 0,$$

which implies that $\widetilde{r}_{N+1}(T)$ decreases strictly with T from θ to 0. When $N = 0$, $\widetilde{r}_1(T) \equiv \theta e^{-\theta T}$. Similarly, for $0 < T < \infty$,

$$\lim_{N \to 0} \widetilde{r}_{N+1}(T) = \theta e^{-\theta T}, \qquad \lim_{N \to \infty} \widetilde{r}_{N+1}(T) = \theta,$$

$$\tilde{r}_{N+1}(T) - \tilde{r}_N(T) = \frac{\theta}{\sum_{j=N}^{\infty}[(\theta T)^j]/j! \sum_{j=N-1}^{\infty}[(\theta T)^j]/j!}$$

$$\times \frac{(\theta T)^{N-1}}{N!} \sum_{j=N}^{\infty} \frac{(\theta T)^j}{j!}(j-N) > 0,$$

which implies that $\tilde{r}_{N+1}(T)$ increases strictly with N from $\theta e^{-\theta T}$ to θ. □

(3) For $0 < T \le \infty$ and $N = 0, 1, 2, \ldots$,

$$Q_N(T; \theta) = \frac{\int_0^T (\theta t)^N e^{-\theta t} dF(t)}{\int_0^T (\theta t)^N e^{-\theta t} \overline{F}(t) dt}$$

increases with T from $h(0)$ to $Q_N(\infty; \theta)$, and increases with N from $Q_0(T; \theta)$ to $h(T)$.

Proof First, note that for $0 < T \le \infty$,

$$\lim_{T \to 0} Q_N(T; \theta) = h(0), \quad h(0) \le Q_N(T; \theta) \le h(T).$$

Next, differentiating $Q_N(T; \theta)$ with respect to T,

$$\frac{dQ_N(T; \theta)}{dT} = \frac{(\theta T)^N e^{-\theta T} \overline{F}(T)}{[\int_0^T (\theta t)^N e^{-\theta t} \overline{F}(t) dt]^2} \int_0^T (\theta t)^N e^{-\theta t} \overline{F}(t)[h(T) - h(t)] dt \ge 0,$$

which implies that $Q_N(T; \theta)$ increases with T from $h(0)$ to $Q_N(\infty; \theta)$. Similarly, denote

$$L_1(T) \equiv \int_0^T (\theta t)^{N+1} e^{-\theta t} dF(t) \int_0^T (\theta t)^N e^{-\theta t} \overline{F}(t) dt$$

$$- \int_0^T (\theta t)^N e^{-\theta t} dF(t) \int_0^T (\theta t)^{N+1} e^{-\theta t} \overline{F}(t) dt.$$

Then,

$$L_1(0) = 0,$$

$$L_1'(T) = (\theta T)^N e^{-\theta T} \overline{F}(T) \int_0^T (\theta t)^N e^{-\theta t} \overline{F}(t)(\theta T - \theta t)[h(T) - h(t)] dt \ge 0,$$

which implies that $Q_N(T; \theta)$ increases with N.

Furthermore, for any small δ $(0 < \delta < T)$,

$$Q_N(T;\theta) = \frac{\int_0^{T-\delta}(\theta t)^N e^{-\theta t}\,\mathrm{d}F(t) + \int_{T-\delta}^T (\theta t)^N e^{-\theta t}\,\mathrm{d}F(t)}{\int_0^{T-\delta}(\theta t)^N e^{-\theta t}\overline{F}(t)\mathrm{d}t + \int_{T-\delta}^T (\theta t)^N e^{-\theta t}\overline{F}(t)\mathrm{d}t}$$

$$\geq \frac{h(T-\delta)\int_{T-\delta}^T (\theta t)^N e^{-\theta t}\overline{F}(t)\mathrm{d}t}{\int_0^{T-\delta}(\theta t)^N e^{-\theta t}\overline{F}(t)\mathrm{d}t + \int_{T-\delta}^T (\theta t)^N e^{-\theta t}\overline{F}(t)\mathrm{d}t}$$

$$= \frac{h(T-\delta)}{1 + [\int_0^{T-\delta}(\theta t)^N e^{-\theta t}\overline{F}(t)\mathrm{d}t / \int_{T-\delta}^T (\theta t)^N e^{-\theta t}\overline{F}(t)\mathrm{d}t]}.$$

The quantity in the bracket of the denominator is

$$\frac{\int_0^{T-\delta}(\theta t)^N e^{-\theta t}\overline{F}(t)\mathrm{d}t}{\int_{T-\delta}^T (\theta t)^N e^{-\theta t}\overline{F}(t)\mathrm{d}t} \leq \frac{1}{\delta e^{-\theta T}\overline{F}(T)} \int_0^{T-\delta}\left(\frac{t}{T-\delta}\right)^N \mathrm{d}t \to 0$$

as $N \to \infty$. So that,

$$h(T-\delta) \leq Q_N(T;\theta) \leq h(T),$$

which follows that $\lim_{N\to\infty} Q_N(T;\theta) = h(T)$ because δ is arbitrary. In addition, taking that T goes to ∞,

$$\lim_{N\to\infty} \frac{\int_0^\infty (\theta t)^N e^{-\theta t}\,\mathrm{d}F(t)}{\int_0^\infty (\theta t)^N e^{-\theta t}\overline{F}(t)\mathrm{d}t} = h(\infty). \qquad \Box$$

Next, we investigates the properties of $Q_0(T;\theta)$:

(4) For $0 < T \leq \infty$,

$$Q_0(T;\theta) \equiv \frac{\int_0^T e^{-\theta t}\,\mathrm{d}F(t)}{\int_0^T e^{-\theta t}\overline{F}(t)\mathrm{d}t}$$

increases with T from $h(0)$ to $\theta F^*(\theta)/[1 - F^*(\theta)]$ and decreases with θ from $F(T)/\int_0^T \overline{F}(t)\mathrm{d}t$ to $h(0)$.

Proof It has been already proved in **(3)** that $Q_0(T;\theta)$ increases with T from $h(0)$ to

$$\lim_{T\to\infty} \frac{\int_0^T e^{-\theta t}\,\mathrm{d}F(t)}{\int_0^T e^{-\theta t}\overline{F}(t)\mathrm{d}t} = \frac{\theta F^*(\theta)}{1 - F^*(\theta)},$$

where $F^*(\theta)$ is the LS transform of $F(t)$, i.e. $F^*(\theta) \equiv \int_0^\infty e^{-\theta t} dF(t)$ for $\theta > 0$. Clearly,

$$\lim_{\theta \to 0} \frac{\int_0^T e^{-\theta t} dF(t)}{\int_0^T e^{-\theta t} \overline{F}(t) dt} = \frac{F(T)}{\int_0^T \overline{F}(t) dt}.$$

Differentiating $Q_0(T; \theta)$ with respect to θ,

$$\frac{dQ_0(T; \theta)}{d\theta} = \frac{1}{[\int_0^T e^{-\theta t} \overline{F}(t) dt]^2} \left\{ \int_0^T te^{-\theta t} \overline{F}(t) dt \int_0^T e^{-\theta t} dF(t) \right.$$
$$\left. - \int_0^T te^{-\theta t} dF(t) \int_0^T e^{-\theta t} \overline{F}(t) dt \right\}.$$

Letting $L_2(T)$ be the bracket of the right-hand side,

$$L_2(0) = 0,$$

$$L_2'(T) = e^{-\theta T} \overline{F}(T) \int_0^T e^{-\theta t} \overline{F}(t) [h(t) - h(T)](T - t) dt \le 0,$$

which implies that $L_2(T) \le 0$, i.e. $Q_0(T; \theta)$ decreases with θ. Furthermore,

$$\lim_{\theta \to \infty} \frac{\int_0^T e^{-\theta t} dF(t)}{\int_0^T e^{-\theta t} \overline{F}(t) dt} = \lim_{\theta \to \infty} \frac{\int_0^T f(t) d(1 - e^{-\theta t})}{\int_0^T \overline{F}(t) d(1 - e^{-\theta t})} = \frac{f(0)}{\overline{F}(0)} = h(0),$$

because $\lim_{\theta \to \infty} (1 - e^{-\theta t})$ is the degenerate distribution placing unit mass at $t = 0$. Therefore, $Q_0(T; \theta)$ decreases with θ from $F(T)/\int_0^T \overline{F}(t) dt$ to $h(0)$.

We make the following another proof: For any small δ $(0 < \delta < T)$,

$$\frac{\int_0^T e^{-\theta t} dF(t)}{\int_0^T e^{-\theta t} \overline{F}(t) dt} = \frac{\int_0^\delta e^{-\theta t} dF(t) + \int_\delta^T e^{-\theta t} dF(t)}{\int_0^\delta e^{-\theta t} \overline{F}(t) dt + \int_\delta^T e^{-\theta t} \overline{F}(t) dt}$$

$$\le \frac{h(\delta) \int_0^\delta e^{-\theta t} \overline{F}(t) dt + h(T) \int_\delta^T e^{-\theta t} \overline{F}(t) dt}{\int_0^\delta e^{-\theta t} \overline{F}(t) dt}$$

$$= h(\delta) + h(T) \frac{\int_\delta^T e^{-\theta t} \overline{F}(t) dt}{\int_0^\delta e^{-\theta t} \overline{F}(t) dt}.$$

The fraction of the right-hand side is

$$\frac{\int_\delta^T e^{-\theta t}\overline{F}(t)dt}{\int_0^\delta e^{-\theta t}\overline{F}(t)dt} \le \frac{\int_\delta^T e^{-\theta t}dt}{\delta e^{-\theta\delta}} = \frac{1}{\delta}\int_\delta^T e^{-\theta(t-\delta)}dt \to 0 \quad \text{as } \theta \to \infty.$$

Thus, from $Q_0(T;\theta) \ge h(0)$,

$$h(0) \le \lim_{\theta\to\infty} \frac{\int_0^T e^{-\theta t}dF(t)}{\int_0^T e^{-\theta t}\overline{F}(t)dt} \le h(\delta),$$

which follows that $\lim_{\theta\to\infty} Q_0(T;\theta) = h(0)$ because δ is arbitrary. In addition, taking that T goes to ∞,

$$\lim_{\theta\to\infty} \frac{\int_0^\infty e^{-\theta t}dF(t)}{\int_0^\infty e^{-\theta t}\overline{F}(t)dt} = h(0). \qquad \square$$

We have the following properties of $Q_1(T;\theta)$:

(5) For $0 < T \le \infty$,

$$Q_1(T;\theta) \equiv \frac{\int_0^T te^{-\theta t}dF(t)}{\int_0^T te^{-\theta t}\overline{F}(t)dt}$$

increases with T from $h(0)$ to $Q_1(\infty;\theta)$ and decreases with θ from $Q_1(T;0) = \int_0^T tdF(t)/\int_0^T t\overline{F}(t)dt$ to $h(0)$. Taking that T goes to ∞,

$$\lim_{\theta\to\infty} \frac{\int_0^\infty te^{-\theta t}dF(t)}{\int_0^\infty te^{-\theta t}\overline{F}(t)dt} = h(0).$$

So that, $Q_1(\infty;\theta)$ decreases with θ from $Q_1(\infty;0)$ to $h(0)$.

(6) For $0 \le T < \infty$ and $N = 0, 1, 2, \ldots$,

$$\tilde{Q}_N(T;\theta) = \frac{\int_T^\infty (\theta t)^N e^{-\theta t}dF(t)}{\int_T^\infty (\theta t)^N e^{-\theta t}\overline{F}(t)dt}$$

increases with T from $\tilde{Q}_N(0;\theta) = Q_N(\infty;\theta)$ in **(3)** to $h(\infty)$ and increases with N from $\tilde{Q}_0(T;\theta) = \int_T^\infty e^{-\theta t}dF(t)/\int_T^\infty e^{-\theta t}\overline{F}(t)dt$ to $h(\infty)$.

Proof First, note that for $0 \le T < \infty$,

$$\lim_{T \to \infty} \tilde{Q}_N(T; \theta) = h(\infty), \quad h(T) \le \tilde{Q}_N(T; \theta) \le h(\infty).$$

Next, differentiating $\tilde{Q}_N(T; \theta)$ with respect to T,

$$\frac{d\tilde{Q}_N(T; \theta)}{dT} = \frac{(\theta T)^N e^{-\theta T} \overline{F}(T)}{[\int_T^\infty (\theta t)^N e^{-\theta t} \overline{F}(t) dt]^2} \int_T^\infty (\theta t)^N e^{-\theta t} \overline{F}(t)[h(t) - h(T)] dt \ge 0,$$

which implies that $\tilde{Q}_N(T; \theta)$ increases with T from $\tilde{Q}_N(0; \theta)$ to $h(\infty)$.
    Similarly, denote

$$L_3(T) \equiv \int_T^\infty (\theta t)^{N+1} e^{-\theta t} dF(t) \int_T^\infty (\theta t)^N e^{-\theta t} \overline{F}(t) dt$$

$$- \int_T^\infty (\theta t)^N e^{-\theta t} dF(t) \int_T^\infty (\theta t)^{N+1} e^{-\theta t} \overline{F}(t) dt.$$

Then,

$$L_3(\infty) = 0,$$

$$L_3'(T) = (\theta T)^N e^{-\theta T} \overline{F}(T) \int_T^\infty (\theta t)^N e^{-\theta t} \overline{F}(t)(\theta t - \theta T)[h(T) - h(t)] dt \le 0,$$

which implies that $L_3'(T) > 0$, i.e. $\tilde{Q}_N(T; \theta)$ increases with N from $\tilde{Q}_0(T; \theta)$.
Furthermore, for any small δ $(0 < \delta < T)$ and any large $T_1 > T$,

$$h(T) \le \frac{\int_T^{T_1} (\theta t)^N e^{-\theta t} dF(t)}{\int_T^{T_1} (\theta t)^N e^{-\theta t} \overline{F}(t) dt} \le h(T_1),$$

and

$$\frac{\int_T^{T_1} (\theta t)^N e^{-\theta t} dF(t)}{\int_T^{T_1} (\theta t)^N e^{-\theta t} \overline{F}(t) dt} = \frac{\int_T^{T_1-\delta} (\theta t)^N e^{-\theta t} dF(t) + \int_{T_1-\delta}^{T_1} (\theta t)^N e^{-\theta t} dF(t)}{\int_T^{T_1-\delta} (\theta t)^N e^{-\theta t} \overline{F}(t) dt + \int_{T_1-\delta}^{T_1} (\theta t)^N e^{-\theta t} \overline{F}(t) dt}$$

$$\ge \frac{h(T_1 - \delta) \int_{T_1-\delta}^{T_1} (\theta t)^N e^{-\theta t} \overline{F}(t) dt}{\int_T^{T_1-\delta} (\theta t)^N e^{-\theta t} \overline{F}(t) dt + \int_{T_1-\delta}^{T_1} (\theta t)^N e^{-\theta t} \overline{F}(t) dt}$$

$$= \frac{h(T_1 - \delta)}{1 + [\int_T^{T_1-\delta} (\theta t)^N e^{-\theta t} \overline{F}(t) dt / \int_{T_1-\delta}^{T_1} (\theta t)^N e^{-\theta t} \overline{F}(t) dt]}.$$

The quantity in the bracket of the denominator is

$$\frac{\int_T^{T_1-\delta} (\theta t)^N e^{-\theta t} \overline{F}(t) dt}{\int_{T_1-\delta}^{T_1} (\theta t)^N e^{-\theta t} \overline{F}(t) dt} \le \frac{e^{-\theta T} \overline{F}(T)}{\delta e^{-\theta T_1} \overline{F}(T_1)} \int_T^{T_1-\delta} \left(\frac{t}{T_1 - \delta} \right)^N dt \to 0 \quad \text{as} \quad N \to 0.$$

Thus,

$$h(T_1 - \delta) \le \lim_{N \to \infty} \frac{\int_T^{T_1} (\theta t)^N e^{-\theta t} dF(t)}{\int_T^{T_1} (\theta t)^N e^{-\theta t} \overline{F}(t) dt} \le h(T_1),$$

which follows that because δ is arbitrary,

$$\lim_{N \to \infty} \frac{\int_T^{T_1} (\theta t)^N e^{-\theta t} dF(t)}{\int_T^{T_1} (\theta t)^N e^{-\theta t} \overline{F}(t) dt} = h(T_1).$$

Furthermore, because T_1 is also arbitrary, $\lim_{N \to \infty} \tilde{Q}_N(T; \theta) = h(\infty)$.   □

Next, we investigate the properties of $\tilde{Q}_0(T; \theta)$ and $\tilde{Q}(T; \theta)$:

(7) For $0 \le T < \infty$,

$$\tilde{Q}_0(T; \theta) = \frac{\int_T^\infty e^{-\theta t} dF(t)}{\int_T^\infty e^{-\theta t} \overline{F}(t) dt}$$

increases with T from $\tilde{Q}_0(0; \theta) = Q_0(\infty; \theta) = \theta F^*(\theta)/[1 - F^*(\theta)]$ in (4) to $h(\infty)$, and decreases with θ from $\tilde{Q}_0(T; 0) = \overline{F}(T)/\int_T^\infty \overline{F}(t) dt$ to $h(T)$.

Proof It has been already proved in (6) that $\tilde{Q}_0(T; \theta)$ increases with T from $\tilde{Q}_0(0; \theta)$ to $h(\infty)$. Clearly,

$$\lim_{\theta \to 0} \frac{\int_T^\infty e^{-\theta t} dF(t)}{\int_T^\infty e^{-\theta t} \overline{F}(t) dt} = \frac{\overline{F}(T)}{\int_T^\infty \overline{F}(t) dt}.$$

Differentiating $\tilde{Q}_0(T;\theta)$ with respect to θ,

$$\frac{d\tilde{Q}_0(T;\theta)}{d\theta} = \frac{1}{[\int_T^\infty e^{-\theta t}\overline{F}(t)dt]^2} \left\{ \int_T^\infty te^{-\theta t}\overline{F}(t)dt \int_T^\infty e^{-\theta t}dF(t) \right.$$
$$\left. - \int_T^\infty te^{-\theta t}dF(t) \int_T^\infty e^{-\theta t}\overline{F}(t)dt \right\}.$$

Letting $L_4(T)$ be the bracket of the right-hand side,

$$L_4(\infty) = 0,$$

$$L_4'(T) = e^{-\theta T}\overline{F}(T) \int_T^\infty e^{-\theta t}\overline{F}(t)[h(t) - h(T)](t - T)dt \geq 0,$$

which implies that $L_4(T) \leq 0$, i.e. $\tilde{Q}_0(T;\theta)$ decreases with θ from $\overline{F}(T)/\int_T^\infty \overline{F}(t)dt$. Furthermore, by the similar method used in **(3)**,

$$\lim_{\theta\to\infty} \frac{\int_T^\infty e^{-\theta t}dF(t)}{\int_T^\infty e^{-\theta t}\overline{F}(t)dt} = \lim_{\theta\to\infty} \frac{\int_T^\infty \theta e^{-\theta(t-T)}f(t)dt}{\int_T^\infty \theta e^{-\theta(t-T)}\overline{F}(t)dt}$$
$$= \lim_{\theta\to\infty} \frac{\int_0^\infty f(t+T)d(1-e^{-\theta t})}{\int_0^\infty \overline{F}(t+T)d(1-e^{-\theta t})} = h(T).$$

Therefore, $\tilde{Q}_0(T;\theta)$ decreases with θ from $\overline{F}(T)/\int_T^\infty \overline{F}(t)dt$ to $h(T)$.
We make the following another proof: For any small $\delta > 0$,

$$\frac{\int_T^\infty e^{-\theta t}dF(t)}{\int_T^\infty e^{-\theta t}\overline{F}(t)dt} = \frac{\int_T^{T+\delta} e^{-\theta t}dF(t) + \int_{T+\delta}^\infty e^{-\theta t}dF(t)}{\int_T^{T+\delta} e^{-\theta t}\overline{F}(t)dt + \int_{T+\delta}^\infty e^{-\theta t}\overline{F}(t)dt}$$
$$\leq \frac{h(T+\delta)\int_T^{T+\delta} e^{-\theta t}\overline{F}(t)dt + \int_{T+\delta}^\infty e^{-\theta t}dF(t)}{\int_T^{T+\delta} e^{-\theta t}\overline{F}(t)dt}$$
$$= h(T+\delta) + \frac{\int_{T+\delta}^\infty e^{-\theta t}dF(t)}{\int_T^{T+\delta} e^{-\theta t}\overline{F}(t)dt}.$$

The fraction of the right-hand side is

$$\frac{\int_{T+\delta}^{\infty} e^{-\theta t} dF(t)}{\int_{T}^{T+\delta} e^{-\theta t} \overline{F}(t) dt} \le \frac{\int_{T+\delta}^{\infty} e^{-\theta t} dF(t)}{\delta e^{-\theta(t+\delta)} \overline{F}(T+\delta)}$$

$$= \frac{1}{\delta \overline{F}(T+\delta)} \int_{T+\delta}^{\infty} e^{-\theta(t-T-\delta)} dF(t) \to 0 \text{ as } \theta \to \infty.$$

Thus, because $\widetilde{Q}_0(T; \theta) \ge h(T)$,

$$h(T) \le \lim_{\theta \to \infty} \frac{\int_{T}^{\infty} e^{-\theta t} dF(t)}{\int_{T}^{\infty} e^{-\theta t} \overline{F}(t) dt} \le h(T+\delta),$$

which follows that $\lim_{\theta \to \infty} \widetilde{Q}_0(T; \theta) = h(T)$ because δ is arbitrary. $\square$

(8) For $0 \le T < \infty$,

$$\widetilde{Q}_1(T; \theta) = \frac{\int_{T}^{\infty} t e^{-\theta t} dF(t)}{\int_{T}^{\infty} t e^{-\theta t} \overline{F}(t) dt}$$

increases with T from $\widetilde{Q}_1(0; \theta) = Q_1(\infty; \theta)$ in **(5)** to $h(\infty)$ and decreases with θ from $\widetilde{Q}_1(T; 0)$ to $h(T)$.

We have the properties of $Q_N(T; 0) \equiv Q_N(T)$ and $\widetilde{Q}_N(T; 0) \equiv \widetilde{Q}_N(T)$:

(9) For $0 < T \le \infty$ and $N = 0, 1, 2, \ldots$,

$$Q_N(T) = \frac{\int_{0}^{T} t^N dF(t)}{\int_{0}^{T} t^N \overline{F}(t) dt}$$

increases with T from $h(0)$ to $Q_N(\infty)$, and increases with N from $F(T)/\int_{0}^{T} \overline{F}(t) dt$ to $h(T)$.

(10) For $0 \le T < \infty$ and $N = 0, 1, 2, \ldots$,

$$\widetilde{Q}_N(T) = \frac{\int_{T}^{\infty} t^N dF(t)}{\int_{T}^{\infty} t^N \overline{F}(t) dt}$$

increases with T from $\widetilde{Q}_N(0) = Q_N(\infty)$ in **(9)** to $h(\infty)$, and increases with N from $\overline{F}(T)/\int_{T}^{\infty} \overline{F}(t) dt$ to $h(\infty)$.

From the above results, we have the following inequalities:

(11) For $0 < T < \infty$ and $0 < \theta < \infty$,

$$\frac{\int_0^T (\theta t)^N e^{-\theta t} dF(t)}{\int_0^T (\theta t)^N e^{-\theta t} \overline{F}(t) dt} \leq \frac{\int_0^T t^N dF(t)}{\int_0^T t^N \overline{F}(t) dt} \leq h(T)$$

$$\leq \frac{\int_T^\infty (\theta t)^N e^{-\theta t} dF(t)}{\int_T^\infty (\theta t)^N e^{-\theta t} \overline{F}(t) dt} \leq \frac{\int_T^\infty t^N dF(t)}{\int_T^\infty t^N \overline{F}(t) dt} \quad (N = 0, 1, 2, \ldots),$$

$$\frac{\int_0^T e^{-\theta t} dF(t)}{\int_0^T e^{-\theta t} \overline{F}(t) dt} \leq \frac{F(T)}{\int_0^T \overline{F}(t) dt} \leq \frac{\int_0^T t dF(t)}{\int_0^T t \overline{F}(t) dt} \leq h(T)$$

$$\leq \frac{\int_T^\infty e^{-\theta t} dF(t)}{\int_T^\infty e^{-\theta t} \overline{F}(t) dt} \leq \frac{\overline{F}(T)}{\int_T^\infty \overline{F}(t) dt} \leq \frac{\int_T^\infty t dF(t)}{\int_T^\infty t \overline{F}(t) dt},$$

where all failure rates are equal to $1/\lambda$ when $F(t) = 1 - e^{-\lambda t}$ $(0 < \lambda < \infty)$.

A.2 Properties of Failure Rates $h(t)$

We give the following properties of the failure rate $h(t)$:

(12) For $0 < T \leq \infty$ and $N = 0, 1, 2, \ldots$,

$$H_N(T; \theta) = \frac{\int_0^T (\theta t)^N e^{-\theta t} h(t) dt}{\int_0^T (\theta t)^N e^{-\theta t} dt}$$

increases with T from $h(0)$ to $H_N(\infty; \theta)$ and increases with N from $H_0(T; \theta) = \int_0^T \theta e^{-\theta t} h(t) dt / (1 - e^{-\theta T})$ to $h(T)$.

(13) For $0 < T \leq \infty$,

$$H_0(T; \theta) = \frac{\int_0^T e^{-\theta t} h(t) dt}{\int_0^T e^{-\theta t} dt}$$

increases with T from $h(0)$ to $\int_0^\infty \theta e^{-\theta t} h(t) dt$ and decreases with θ from $H(T)/T$ to $h(0)$.

(14) For $0 < T \leq \infty$ and $N = 0, 1, 2, \ldots$,

$$H_N(T) = \frac{\int_0^T t^N h(t) \, dt}{\int_0^T t^N \, dt}$$

increases with T from $h(0)$ to $h(\infty)$, and increases with N from $H(T)/T$ to $h(T)$.

(15) For $0 \le T < \infty$ and $N = 0, 1, 2, \ldots$,

$$\tilde{H}_N(T; \theta) = \frac{\int_T^\infty (\theta t)^N e^{-\theta t} h(t) \, dt}{\int_T^\infty (\theta t)^N e^{-\theta t} \, dt}$$

increases with T from $\tilde{H}_N(0; \theta) = H_N(\infty; \theta)$ in **(12)** to $h(\infty)$ and increases with N from $\tilde{H}_0(T; \theta)$ to $h(\infty)$.

(16) For $0 \le T < \infty$,

$$\tilde{H}_0(T; \theta) = \frac{\int_T^\infty e^{-\theta t} h(t) \, dt}{\int_T^\infty e^{-\theta t} \, dt}$$

increases with T from $\int_0^\infty \theta e^{-\theta t} h(t) \, dt$ to $h(\infty)$ and decreases with θ from $h(\infty)$ to $h(T)$.

Proof $\tilde{H}_0(T; \theta)$ is rewritten as

$$\tilde{H}_0(T; \theta) = \int_0^\infty h(t + T) \, d(1 - e^{-\theta t}).$$

When $\theta \to 0$ and $\theta \to \infty$, $1 - e^{-\theta t}$ is a degenerate distribution placing unit mass at $t = \infty$ and $t = 0$, respectively, which can complete the proof. □

(17) For $0 < \theta < \infty$ and $0 < T < \infty$,

$$\frac{\int_0^T (\theta t)^N e^{-\theta t} h(t) \, dt}{\int_0^T (\theta t)^N e^{-\theta t} \, dt} \le \frac{\int_0^T t^N h(t) \, dt}{\int_0^T t^N \, dt} \le h(T) \le \frac{\int_T^\infty (\theta t)^N e^{-\theta t} h(t) \, dt}{\int_T^\infty (\theta t)^N e^{-\theta t} \, dt}$$

$$(N = 0, 1, 2, \ldots),$$

$$\frac{\int_0^T e^{-\theta t} \, dF(t)}{\int_0^T e^{-\theta t} \overline{F}(t) \, dt} \le \frac{\int_0^T e^{-\theta t} h(t) \, dt}{\int_0^T e^{-\theta t} \, dt} \le \frac{H(T)}{T} \le \frac{\int_0^T t h(t) \, dt}{\int_0^T t \, dt} \le h(T)$$

$$\le \frac{\int_T^\infty e^{-\theta t} \, dF(t)}{\int_T^\infty e^{-\theta t} \overline{F}(t) \, dt} \le \frac{\int_T^\infty e^{-\theta t} h(t) \, dt}{\int_T^\infty e^{-\theta t} \, dt},$$

where all failures rates are equal to $1/\lambda$ when $F(t) = 1 - e^{-\lambda t}$ $(0 < \lambda < \infty)$.

Proof Prove that

$$\frac{\int_0^T e^{-\theta t} h(t) dt}{\int_0^T e^{-\theta t} dt} \geq \frac{\int_0^T e^{-\theta t} dF(t)}{\int_0^T e^{-\theta t} \overline{F}(t) dt}.$$

Denote that

$$L_5(T) \equiv \int_0^T e^{-\theta t} h(t) dt \int_0^T e^{-\theta t} \overline{F}(t) dt - \int_0^T e^{-\theta t} dt \int_0^T e^{-\theta t} dF(t).$$

Then,

$$L_5(0) = 0,$$

$$L_5'(T) = e^{-\theta T} h(T) \int_0^T e^{-\theta t} \overline{F}(t) dt + e^{-\theta T} \overline{F}(T) \int_0^T e^{-\theta t} h(t) dt$$

$$- e^{-\theta T} \int_0^T e^{-\theta t} dF(t) - e^{-\theta T} f(T) \int_0^T e^{-\theta t} dt$$

$$= e^{-\theta T} \int_0^T e^{-\theta t} [h(T) - h(t)][F(T) - F(t)] dt \geq 0,$$

which implies that $L_5(T) \geq 0$, and completes the proof. Similarly, the other inequalities have been already derived in **(3)–(16)** or are easily proved.   □

Appendix B
Answers to Selected Problems

Chapter 1

* 1.2 For example, consider consecutive K-out-of-n systems when K is a random variable, as shown in Sect. 6.3.2.

* 1.3 Read Sect. 8.4 and consider other random reliability models.

1.5 Use that $0 < \Gamma(\alpha) < 1$ for $1 < \alpha < 2$ and $\Gamma(\alpha) \geq 1$ for the others.

1.6 For $0 < t < \infty$,

$$\frac{\lambda}{\lambda+\mu}[1 - e^{-(\lambda+\mu)t}] - \frac{\lambda t}{1+(\lambda+\mu)t}$$
$$= \frac{\lambda}{(\lambda+\mu)[1+(\lambda+\mu)t]}\{1 - [1+(\lambda+\mu)t]e^{-(\lambda+\mu)t}\} > 0.$$

1.7 For $0 < t < \infty$,

$$\frac{\lambda^2 t^2}{1+2\lambda t} - \frac{\lambda t}{2} + \frac{1}{4}(1 - e^{-2\lambda t}) = \frac{1}{4(1+2\lambda t)}[1 - (1+2\lambda t)e^{-2\lambda t}] > 0.$$

1.8 The LS transform of $F(t)$ is

$$F^*(s) \equiv \int_0^\infty e^{-st} dF(t) = \int_0^\infty e^{-st} f(t) dt = -\int_0^\infty f(t) d\left(e^{-st}/s\right)$$

$$= -f(t)\frac{1}{s}e^{-st}\Big|_0^\infty + \int_0^\infty \frac{1}{s}e^{-st} df(t) = \frac{1}{s}f^*(s).$$

1.9 For $0 < t < \infty$,

© Springer-Verlag London 2014
T. Nakagawa, *Random Maintenance Policies*,
Springer Series in Reliability Engineering, DOI 10.1007/978-1-4471-6575-0

$$\left(\frac{\lambda}{\lambda+\mu}\right)^2 [1 - e^{-(\lambda+\mu)t}] + \frac{\lambda\mu t}{\lambda+\mu} - \frac{\lambda t(1+\mu t)}{1+(\lambda+\mu)t}$$

$$= \frac{\lambda^2}{(\lambda+\mu)^2[1+(\lambda+\mu)t]}\{1 - [1+(\lambda+\mu)t]e^{-(\lambda+\mu)t}\} > 0.$$

1.10 See [11, p.19, p.200].

Chapter 2

2.1 Use that

$$\int_0^\infty \overline{G}(t)dF(t) = \int_0^\infty F(t)dG(t), \quad \int_0^\infty G(t)dF(t) = \int_0^\infty \overline{F}(t)dG(t),$$

$$\int_0^\infty \overline{G}(t)\overline{F}(t)dt = \int_0^\infty \left[\int_0^t \overline{F}(u)du\right]dG(t) = \int_0^\infty \left[\int_0^t \overline{G}(u)du\right]dF(t).$$

2.2 Setting $C_1'(\theta) = 0$ in (2.8),

$$\int_0^\infty \theta^2 t e^{-\theta t}\overline{F}(t)dt = \int_0^\infty \overline{F}(t)d[1 - (1+\theta t)e^{-\theta t}]$$

$$= \int_0^\infty [1 - (1+\theta t)e^{-\theta t}]dF(t) = \frac{c_E}{c_S + c_E}.$$

2.3 From $[\theta/(\theta+\lambda)]^2 = c_E/(c_S + c_E)$,

$$\frac{[\theta/(\theta+\lambda)]^2}{1 - [\theta/(\theta+\lambda)]^2} = \frac{c_E}{c_S}.$$

So that,

$$\left(\frac{\theta}{\lambda}\right)^2 - \frac{[\theta/(\theta+\lambda)]^2}{1 - [\theta/(\theta+\lambda)]^2} = \frac{2\theta^3}{\lambda^2(2\theta+\lambda)} > 0.$$

2.4 The left-hand side of (2.29) decreases with θ from $\int_0^T \overline{F}(t)[h(T) - h(t)]dt$, which agrees with that of (2.3), and hence, T_F^* increases with θ from T_S^*.

2.5 The left-hand side of (2.39) when $G(t) = 1 - e^{-\theta t}$ increases with θ to $\int_0^T \overline{F}(t)[h(T) - h(t)]dt$ which agrees with that of (2.3), and hence, T_L^* decreases with θ to T_S^*.

2.6 (2.42) + (2.43) + (2.44) is

$$\sum_{j=0}^{\infty} \int_0^T \left[\int_{T-t}^{\infty} \overline{F}(T)dG(u) \right] dG^{(j)}(t) + F(T)$$

$$= \overline{F}(T) \sum_{j=0}^{\infty} [G^{(j)}(T) - G^{(j+1)}(T)] + F(T) = 1.$$

2.7 Differentiating the left-hand side of (2.48) with respect to T,

$$\widetilde{Q}_0'(T; \theta) \int_0^T \overline{F}(t)dt$$

$$+ \frac{1}{\int_T^{\infty} e^{-\theta t} \overline{F}(t)dt} \left\{ \overline{F}(T) \int_T^{\infty} e^{-\theta t} \overline{F}(t)[h(t) - h(T)]dt \right\} > 0.$$

Thus, because $\widetilde{Q}_0(T; \theta)$ increases strictly to $h(\infty)$, if $h(\infty) > c_F/[\mu(c_F - c_R)]$, then there exists a finite and unique T_O^* which satisfies (2.48), and the resulting cost rate is given in (2.49).

2.9 Because $T_O^* < T_F^*$, $C_F(T_F^*) = (c_F - c_T)h(T_F^*)$ and $C_O(T_O^*) = (c_F - c_T)\widetilde{Q}_0(T_O^*; \theta)$ when $c_T = c_R$, if $\widetilde{Q}_0(T_O^*; \theta) < h(T_F^*)$, then $C_O(T_O^*) < C_F(T_F^*)$. Similarly, if $\widetilde{Q}_0(T_O^*; \theta) < h(T_L^*)$, then $C_O(T_O^*) < C_L(T_L^*)$.

2.10 Let $L_1(T)$ be

$$L_1(T) \equiv \int_0^T t^{\alpha_1} dF(t) \int_0^T t^{\alpha_2} \overline{F}(t)dt - \int_0^T t^{\alpha_2} dF(t) \int_0^T t^{\alpha_1} \overline{F}(t)dt.$$

Then, for $\alpha_1 > \alpha_2 > 0$,

$$L_1(0) = 0,$$

$$L_1'(T) = T^{\alpha_2} \overline{F}(T) \int_0^T t^{\alpha_2} \overline{F}(t)(T^{\alpha_1-\alpha_2} - t^{\alpha_1-\alpha_2})[h(T) - h(t)]dt > 0,$$

which implies that $L_1(T) > 0$ for $0 < T < \infty$. Thus, $\int_0^T t^{\alpha} dF(t)/\int_0^T t^{\alpha} \overline{F}(t)dt$ increases strictly with α. Similarly, let $L_2(T)$ be

$$L_2(T) \equiv \int_T^\infty t^{\alpha_1} dF(t) \int_T^\infty t^{\alpha_2} \overline{F}(t) dt - \int_T^\infty t^{\alpha_2} dF(t) \int_T^\infty t^{\alpha_1} \overline{F}(t) dt.$$

Then,

$$L_2(\infty) = 0,$$

$$L_2'(T) = T^{\alpha_2} \overline{F}(T) \int_T^\infty t^{\alpha_2} \overline{F}(t)(t^{\alpha_1 - \alpha_2} - T^{\alpha_1 - \alpha_2})[h(T) - h(t)] dt < 0,$$

which implies that $L_2(T) > 0$ for $0 < T < \infty$. Thus, $\int_T^\infty t^\alpha dF(t)/\int_T^\infty \overline{F}(t) dt$ increases strictly with α.

2.12 The probability that the unit is replaced at number N before time T is

$$\int_0^T \overline{F}(t) dG^{(N)}(t),$$

the probability that it is replaced at the first completion of working times over time T before number N is

$$\sum_{j=0}^{N-1} \int_0^T \left[\int_{T-t}^\infty \overline{F}(t+u) dG(u) \right] dG^{(j)}(t),$$

the probability that it is replaced at failure before time T is

$$\int_0^T [1 - G^{(N)}(t)] dF(t),$$

and the probability that it is replaced at failure after time T is

$$\sum_{j=0}^{N-1} \int_0^T \left\{ \int_{T-t}^\infty [F(t+u) - F(T)] dG(u) \right\} dG^{(j)}(t).$$

The mean time to replacement is

$$
\int_0^T t\overline{F}(t)\mathrm{d}G^{(N)}(t) + \int_0^T t[1 - G^{(N)}(t)]\mathrm{d}F(t)
$$

$$
+ \sum_{j=0}^{N-1} \int_0^T \left[\int_{T-t}^{\infty} (t+u)\overline{F}(t+u)\mathrm{d}G(u) \right] \mathrm{d}G^{(j)}(t)
$$

$$
+ \sum_{j=0}^{N-1} \int_0^T \left\{ \int_{T-t}^{\infty} \left[\int_T^{t+u} y\mathrm{d}F(y) \right] \mathrm{d}G(u) \right\} \mathrm{d}G^{(j)}(t)
$$

$$
= \int_0^T \overline{F}(t)[1 - G^{(N)}(t)]\mathrm{d}t + \sum_{j=0}^{N-1} \int_0^T \left[\int_T^{\infty} \overline{F}(u)\overline{G}(u-t)\,\mathrm{d}u \right] \mathrm{d}G^{(j)}(t).
$$

Therefore, the expected cost rate is

$$
C_O(T, N) = \frac{\begin{aligned}&c_F - (c_F - c_R)\int_0^T \overline{F}(t)\mathrm{d}G^{(N)}(t)\\&-(c_F - c_T)\sum_{j=0}^{N-1}\int_0^T\left[\int_{T-t}^{\infty}\overline{F}(t+u)\mathrm{d}G(u)\right]\mathrm{d}G^{(j)}(t)\end{aligned}}{\begin{aligned}&\int_0^T \overline{F}(t)[1 - G^{(N)}(t)]\mathrm{d}t\\&+\sum_{j=0}^{N-1}\int_0^T\left[\int_T^{\infty}\overline{F}(u)\overline{G}(u-t)\mathrm{d}u\right]\mathrm{d}G^{(j)}(t)\end{aligned}},
$$

where c_F = replacement cost at failure, c_R = replacement cost at number N, and c_T = replacement cost over time T.

2.13 The probability that the unit is replaced at time T after number N is

$$
\overline{F}(T)G^{(N)}(T),
$$

the probability that it is replaced at the first completion of working times over times T before number N is

$$
\sum_{j=0}^{N-1} \int_0^T \left[\int_{T-t}^{\infty} \overline{F}(t+u)\mathrm{d}G(u) \right] \mathrm{d}G^{(j)}(t),
$$

the probability that it is replaced at failure before time T is

$$F(T),$$

and the probability that it is replaced at failure after time T is

$$\sum_{j=0}^{N-1} \int_0^T \left\{ \int_{T-t}^{\infty} [F(t+u) - F(T)] dG(u) \right\} dG^{(j)}(t).$$

The mean time to replacement is

$$T\overline{F}(T)G^{(N)}(T) + \int_0^T t\, dF(t)$$

$$+ \sum_{j=0}^{N-1} \int_0^T \left[\int_{T-t}^{\infty} (t+u)\overline{F}(t+u) dG(u) \right] dG^{(j)}(t)$$

$$+ \sum_{j=0}^{N-1} \int_0^T \left[\int_{T-t}^{\infty} \left[\int_T^{t+u} y\, dF(y) \right] dG(u) \right] dG^{(j)}(t)$$

$$= \int_0^T \overline{F}(t) dt + \sum_{j=0}^{N-1} \int_0^T \left[\int_T^{\infty} \overline{F}(u)\overline{G}(u-t) du \right] dG^{(j)}(t).$$

Therefore, the expected cost rate is

$$C_O(T, N) = \frac{\begin{array}{c} c_F - (c_F - c_R)\overline{F}(T)G^{(N)}(T) \\ -(c_F - c_T)\sum_{j=0}^{N-1} \int_0^T \left[\int_{T-t}^{\infty} \overline{F}(t+u) dG(u) \right] dG^{(j)}(t) \end{array}}{\int_0^T \overline{F}(t) dt + \sum_{j=0}^{N-1} \int_0^T \left[\int_T^{\infty} \overline{F}(u)\overline{G}(u-t) du \right] dG^{(j)}(t)},$$

where c_F, c_R and c_T are given in 2.12.

Chapter 3

3.1 Making integration by parts,

$$\int_0^T t\,dh(t) = Th(T) - H(T) = \int_0^T [h(T) - h(t)]dt.$$

3.2 We can write (3.15) as

$$C_B(G) = \frac{\int_0^\infty Q(t)dG(t)}{\int_0^\infty S(t)dG(t)},$$

where

$$Q(t) \equiv c_F M(t) + c_R, \quad S(t) \equiv t.$$

If there exists a minimum value T $(0 < T \le \infty)$ of $Q(t)/S(t)$, then

$$\frac{Q(t)}{S(t)} \ge \frac{Q(T)}{S(T)},$$

and

$$\int_0^\infty Q(t)dG(t) \ge \frac{Q(T)}{S(T)} \int_0^\infty S(t)dG(t).$$

So that,

$$C_B(G) \ge \frac{Q(T)}{S(T)} = C_B(G_T).$$

3.4 Set that

$$Q(t) \equiv c_D F(t) + c_R, \quad S(t) = t.$$

3.5 For $T > T_1$,

$$\int_0^T \overline{G}(t)[h(T) - h(t)]dt \geq h(T) \int_0^{T_1} \overline{G}(t)dt - \int_0^{T_1} \overline{G}(t)h(t)dt \to \infty$$

$$\text{as} \quad T \to \infty.$$

3.6 Making integration by parts,

$$\int_T^\infty H(u)dG(u - t) = H(T)\overline{G}(T - t) + \int_T^\infty \overline{G}(u - t)h(u)du,$$

and from $M(t) \equiv \sum_{j=1}^\infty G^{(j)}(t)$ and $G^{(0)}(t) \equiv 1$ for $t \geq 0$,

$$\sum_{j=0}^\infty \int_0^T \left[H(T)\overline{G}(T - t) + \int_T^\infty \overline{G}(u - t)h(u)du \right] dG^{(j)}(t)$$

$$= H(T) + \int_T^\infty \overline{G}(t)h(t)dt + \int_0^T \left[\int_T^\infty \overline{G}(u - t)h(u)du \right] dM(t).$$

3.7 The left-hand side of (3.43) increases strictly with T from 0. Furthermore,

$$Th(T) - H(T) \leq Th(T + t) - H(T) \leq (T + t)h(T + t) - H(T + t),$$

i.e.,

$$\int_0^T udh(u) \leq Th(T + t) - H(T) \leq \int_0^{T+t} udh(u).$$

So that, for $0 < t < \infty$,

$$Th(T + t) - H(T) \to \int_0^\infty udh(u) \quad \text{as} \quad T \to \infty.$$

3.8 Use that

$$\int_0^\infty \theta e^{-\theta t} h(t+T) dt = -\int_0^\infty h(t+T) d(e^{-\theta t})$$

$$= h(T) + \int_0^\infty e^{-\theta t} dh(t+T),$$

which decreases with θ to $h(T)$.

3.10 Differentiating $Q(T)$ with θ,

$$\int_0^T t e^{-\theta t} [h(T) - h(t)] dt + \int_T^\infty t e^{-\theta t} [h(t) - h(T)] dt > 0.$$

So that, $Q(T)$ increases with θ from $-\int_T^\infty [h(t) - h(T)] dt$ to $\int_0^T [h(T) - h(t)] dt$, i.e. T_P^* decreases with θ from ∞ to 0.

3.11 Because

$$\frac{N+1}{\theta} H_{N+1} - \int_0^\infty [1 - G^{(N+1)}(t)] h(t) dt - \frac{N}{\theta} H_N$$

$$+ \int_0^\infty [1 - G^{(N)}(t)] h(t) dt = \frac{N+1}{\theta} [H_{N+1} - H_N] > 0,$$

which implies that the left-hand side of (3.54) increases strictly with N. Furthermore, from (12) of Appendix A.2, H_N increases strictly with N to $h(\infty)$ when $G(t) = 1 - e^{-\theta t}$.

3.12 The mean time to replacement is

$$T G^{(N)}(T) + \sum_{j=0}^{N-1} \int_0^T \left[\int_T^\infty u \, dG(u-t) \right] dG^{(j)}(t)$$

$$= T + \sum_{j=0}^{N-1} \int_0^T \left[\int_T^\infty \overline{G}(u-t) du \right] dG^{(j)}(t),$$

and the expected number of failures before replacement is

$$H(T)G^{(N)}(T) + \sum_{j=0}^{N-1} \int_0^T \left[\int_T^\infty H(u) dG(u-t) \right] dG^{(j)}(t)$$

$$= H(T) + \sum_{j=0}^{N-1} \int_0^T \left[\int_T^\infty \overline{G}(u-t) h(u) du \right] dG^{(j)}(t).$$

Therefore, the expected cost rate is

$$\tilde{C}_{OF}(T, N) = \frac{c_M\{\sum_{j=0}^{N-1}\int_0^T \left[\int_T^\infty \overline{G}(u-t)h(u)du\right] dG^{(j)}(t) + H(T)\} + c_R + (c_T - c_R)G^{(N)}(T)}{T + \sum_{j=0}^{N-1}\int_0^T \left[\int_T^\infty \overline{G}(u-t)du\right] dG^{(j)}(t)}.$$

Chapter 4

4.1 From (4.3),

$$\lim_{T\to\infty} C(T)$$

$$= \int_0^\infty \left(\sum_{j=0}^\infty \int_0^t \left\{ \int_{t-x}^\infty [(j+1)c_R + c_D(x+y-t)] dG(y) \right\} dG^{(j)}(x) \right) dF(t),$$

$$\int_0^\infty \left[\int_0^\infty \overline{F}(t+x)\overline{G}(t) dt \right] dM(x) = \int_0^\infty \left[\int_x^\infty \overline{F}(t)\overline{G}(t-x)dt \right] dM(x)$$

$$= \int_0^\infty \left[\int_0^t \overline{G}(t-x)dM(x) \right] \overline{F}(t)dt = \int_0^\infty \overline{F}(t)G(t)dt.$$

4.2 Substitute $G(t) = 1 - e^{-\theta t}$ and $M(t) = \theta t$ for (4.3), and derive (4.5).
4.3 From $T_S^* < \sqrt{2c}$ and $1/\theta^* = \sqrt{c}$, $T_S^* < \sqrt{2}/\theta^*$.
4.4 Differentiate

$$\left[\frac{\theta(e^{\lambda T} - 1) - \lambda(1 - e^{-\theta T})}{\theta(\theta + \lambda)} \right]$$

with θ,

$$\frac{\lambda}{(\theta+\lambda)^2}\left\{\frac{\theta+\lambda}{\theta^2}[1-(1+\theta T)e^{-\theta T}]+\frac{1-e^{-\theta T}}{\theta}-\frac{e^{\lambda T}-1}{\lambda}\right\}<0.$$

Because, the brancket is 0 for $T=0$ and differentiating it with respect to T,

$$[1+(\theta+\lambda)T]e^{-\theta T}-e^{\lambda T}<0.$$

4.6 From (4.52) and (4.63),

$$\frac{1}{\theta}(e^{\lambda T}-1)+\frac{\lambda}{\theta(\theta+\lambda)}e^{-\theta T}>0.$$

4.7 The term of cost c_R in (4.66) is

$$\int_0^S M(t)\mathrm{d}F(t)+\overline{F}(S)M(S)$$

$$+\int_0^S\left\{\sum_{j=0}^\infty\int_0^t[G(S-x)-G(t-x)]\mathrm{d}G^{(j)}(x)\right\}\mathrm{d}F(t)$$

$$=\overline{F}(S)M(S)+\int_0^S\left[\sum_{j=0}^\infty\int_0^t G(S-x)\mathrm{d}G^{(j)}(x)\right]\mathrm{d}F(t)$$

$$=M(S)-\int_0^S F(t)G(S-x)\mathrm{d}M(t)$$

$$=\int_0^S[1-F(t)G(S-t)]\mathrm{d}M(t),$$

and the term of cost c_D is

$$\int_0^S\left\{\sum_{j=0}^\infty\int_0^t\left[\int_{t-x}^{S-x}\overline{G}(y)\mathrm{d}y\right]\mathrm{d}G^{(j)}(x)\right\}\mathrm{d}F(t)$$

$$=\int_0^S\left[\int_t^S\overline{G}(y)\mathrm{d}y\right]\mathrm{d}F(t)+\int_0^S\left\{\int_0^t\left[\int_{t-x}^{S-x}\overline{G}(y)\mathrm{d}y\right]\mathrm{d}M(x)\right\}\mathrm{d}F(t)$$

$$= \int_0^S F(t)\overline{G}(t)\mathrm{d}t + \int_0^S \left\{ \int_x^S \left[\int_t^S \overline{G}(y-x)\mathrm{d}y \right] \mathrm{d}F(t) \right\} \mathrm{d}M(x)$$

$$= \int_0^S F(t)\overline{G}(t)\mathrm{d}t + \int_0^S \left\{ \int_0^{S-x} [F(t+x) - F(x)]\overline{G}(t)\mathrm{d}t \right\} \mathrm{d}M(x).$$

4.9 Replacing $G(t)$ with $G^{(N)}(t)$ ($N = 0, 1, 2, \ldots$) in (4.66), the total expected cost is

$$C_S(N) = c_R \int_0^S [1 - F(t)G^{(N)}(S-t)]\mathrm{d}M^{(N)}(t)$$

$$+ c_D \left(\int_0^S F(t)[1 - G^{(N)}(t)]\mathrm{d}t \right.$$

$$+ \int_0^S \left\{ \int_0^{S-x} [F(t+x) - F(x)][1 - G^{(N)}(t)]\mathrm{d}t \right\} \mathrm{d}M^{(N)}(x) \right),$$

where

$$M^{(N)}(t) \equiv \sum_{j=1}^{\infty} G^{(jN)}(t).$$

When $G(t) = 1 - \mathrm{e}^{-\theta t}$, derive an optimum N^* which minimizes $C_S(N)$.

Chapter 5

5.1 (5.1) + (5.2) is

$$\sum_{k=0}^{\infty} \int_{kT}^{(k+1)T} \left[\sum_{j=0}^{\infty} \int_0^t \overline{G}(t-x)\mathrm{d}G^{(j)}(x) \right] \mathrm{d}F(t)$$

$$= \sum_{k=0}^{\infty} \int_{kT}^{(k+1)T} \mathrm{d}F(t) = \int_0^{\infty} \mathrm{d}F(t) = 1.$$

5.2 When $G(t) \equiv 0$ for $t \geq 0$, noting that $G^{(0)}(t) \equiv 1$ for $t \geq 0$ and $M(t) \equiv 0$,

$$C(T) = c_T \sum_{k=1}^{\infty} \overline{F}(kT) + c_D \mu - c_D \sum_{k=0}^{\infty} (kT) \int_{kT}^{(k+1)T} dF(t)$$

$$= (c_T - c_D T) \sum_{k=1}^{\infty} \overline{F}(kT) + c_D \mu.$$

5.3 Using the approximations $e^{-\lambda T} \approx 1 - \lambda T + (\lambda T)^2/2$, from (5.7),

$$\tilde{T} = \frac{1}{\lambda} \sqrt{\frac{2c_T}{c_D/\lambda}},$$

and when $c_T = 2c_R$, from (5.10),

$$\frac{\lambda}{\theta + \lambda} = \frac{\lambda \tilde{T}}{2}.$$

So that,

$$\frac{C_R(\theta^*)}{c_D/\lambda} = \lambda \tilde{T} - \frac{(\lambda \tilde{T})^2}{4},$$

$$\frac{C_P(T^*)}{c_D/\lambda} = 1 - e^{-\lambda T^*} \approx \lambda T^* - \frac{(\lambda T^*)^2}{2} \approx \frac{C_R(\theta^*)}{c_D/\lambda}.$$

5.4 Differentiating the left-hand side of (5.14) with respect to θ,

$$L(T) \equiv -\frac{1 - (1 + \theta T)e^{-\theta T}}{\theta^2} + \frac{1 - [1 + (\theta + \lambda)T]e^{-(\theta+\lambda)T}}{(\theta + \lambda)^2}.$$

Then,

$$L(0) = 0, \qquad L(\infty) < 0,$$
$$L'(T) = -Te^{-\theta T}(1 - e^{-\lambda T}) < 0,$$

which implies that the left-hand side increases with $1/\theta$, i.e. T_P^* decreases with $1/\theta$ to T^* in (5.7).

5.5 When $F(t) = 1 - e^{-\lambda t}$ and $G(t) = 1 - e^{-\theta t}$,

$$\int_0^\infty \left\{ \int_0^t x[1 - G^{(N)}(t-x)]dG^{(jN)}(x) \right\} dF(t)$$

$$= \int_0^\infty \left\{ \int_x^\infty [1 - G^{(N)}(t-x)]dF(t) \right\} x dG^{(jN)}(x) = \frac{N(1-A)A^N}{\lambda(1-A^N)}.$$

So that,

$$C_{R1}(N) = c_R \frac{A^N}{1-A^N} + \frac{c_D}{\lambda} \left[1 - \frac{N(1-A)A^N}{1-A^N} \right]$$

$$= c_R \frac{A^N}{1-A^N} + \frac{c_D}{\lambda} \frac{(1-A)^2}{1-A^N} \sum_{j=1}^{N} j A^{j-1}.$$

From $C_{R1}(N+1) - C_{R1}(N) \geq 0$,

$$c_R \frac{A^{N+1}}{1-A^{N+1}} + \frac{c_D}{\lambda} \frac{(1-A)^2}{1-A^{N+1}} \sum_{j=1}^{N+1} j A^{j-1}$$

$$- c_R \frac{A^N}{1-A^N} - \frac{c_D}{\lambda} \frac{(1-A)^2}{1-A^N} \sum_{j=1}^{N} j A^{j-1}$$

$$= \frac{(1-A)A^N}{(1-A^N)(1-A^{N+1})} \left[\frac{c_D}{\lambda} (1-A) \sum_{j=1}^{N} (1-A^j) - c_R \right] \geq 0.$$

5.6 Because A decreases with $1/\theta$ from 1 to 0, the left-hand side of (5.26) decreases with A from ∞ to 0, i.e., it increases with $1/\theta$ from 0 to ∞. Thus, N_2^* decreases with $1/\theta$ from ∞ to 1.

5.8 Use (5.23).

5.9 Use

$$\int_0^\infty \left(\int_0^y \left\{ \int_{y-t}^\infty [c_R(j+1) + c_D x]dG(x) \right\} dG^{(j)}(t) \right) dF(y)$$

$$= \int_0^\infty \left\{ \int_0^\infty [c_R(j+1) + c_D x][F(t+x) - F(t)]dG(x) \right\} dG^{(j)}(t).$$

5.10 The bracket of the left-hand side becomes

$$\sum_{j=1}^{N+1} A^j \sum_{j=1}^{N} \left(\frac{1}{A}\right)^j + \sum_{j=1}^{N} \left[\left(\frac{1}{A}\right)^j - 1\right]$$

for $0 < A < 1$, which increases strictly with N to ∞.

5.11 From (5.34) and (5.39),

$$\left(\frac{1-A}{A}\right)^2 \frac{N(N+1)}{A^N} - \frac{(1-A)^2}{A} \left[\frac{(1-A^N)(1-A^{N+2})}{(1-A)^2 A^2} - N\right]$$

$$= \frac{(1-A)^2}{A^{N+2}} \left(N(N+1) - \sum_{j=0}^{N-1} A^{j+1} \sum_{j=0}^{N+1} A^j + NA^{N+1}\right)$$

$$> \frac{N(1-A)^2}{A^{N+2}} \left(N+1 - \sum_{j=0}^{N+1} A^j + A^{N+1}\right) > 0.$$

5.12 From (5.34) and (5.44),

$$\frac{(1-A^N)(1-A^{N+2})}{A^{N+1}} - \frac{N(1-A)^2}{A} - \frac{N(N+1)(1-A)^2}{N(1-A)+1}$$

$$= \frac{(1-A)^2}{A^{N+1}} \left\{\sum_{j=0}^{N-1} A^j \sum_{j=0}^{N+1} A^j - NA^N \left[\frac{N+A+1}{N(1-A)+1}\right]\right\} > 0$$

for $0 < A < 1$, because for $N \geq 1$,

$$\sum_{j=0}^{N-1} A^j > NA^N \quad \text{and} \quad \sum_{j=0}^{N+1} A^j > \frac{N+A+1}{N(1-A)+1}.$$

5.13 From

$$\tilde{L}_3(j) = \int_0^\infty \{e^{-2\lambda t}[c + t + \tilde{L}_3(j+1)] + (1 - e^{-2\lambda t})[c + t + \tilde{L}_3(1)]\}dG(t),$$

$$(j = 1, 2, \ldots, N-1),$$

we have

$$\tilde{L}_3(N) = \tilde{L}_3(1) - \left(c + \frac{1}{\theta}\right) \sum_{j=1}^{N-1} \left[\frac{1}{G^*(2\lambda)}\right]^j,$$

and from (5.50),

$$\widetilde{L}_3(N) = c + \frac{1}{\theta} + c_S G^*(2\lambda) + \widetilde{L}_3(1)[1 - G^*(2\lambda)].$$

Solving two equations for $\widetilde{L}_3(1)$, we get (5.51).

5.14 Consider the following nine cases:

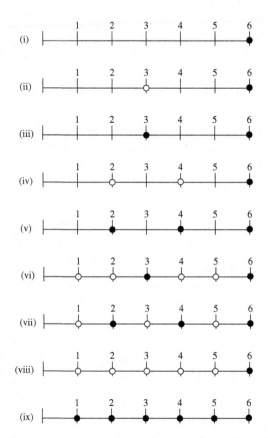

Chapter 6

6.1 Set that $1 - e^{-\lambda t} \equiv x$ and take the integral in (6.2).

6.2 See [16, p.107].

6.3 Differentiate the left-hand side of (6.26) with respect to β and use $\lim_{\beta \to \infty} \int_0^\infty \overline{F}(t) \beta e^{-\beta \overline{F}(t)} dt = 0$.

6.5 Use that

$$1 - \frac{1 - F(t)^{n-1}}{1 - F(t)^n} = \frac{1}{\sum_{j=0}^{n-1}[1/F(t)^j]}.$$

6.6 Letting $L(T; K)$ be the left-hand side of (6.56) when $F(t) = 1 - e^{-\lambda t}$,

$$L(0; K) = 0, \quad L(\infty; K) = K \sum_{j=K}^{n} \frac{1}{j} - 1,$$

$$L'(T; K) = H_n'(T; K) \sum_{j=K}^{n} \binom{n}{j} \int_0^T [\overline{F}(t)]^j [F(t)]^{n-j} dt > 0.$$

Thus, there exists a finite and unique T_1^* which satisfies (6.56).

6.7 Setting that $P_n(\beta) \equiv \sum_{j=0}^{n}(\beta^j/j!)e^{-\beta}$ $(n = 0, 1, 2, \ldots)$,

$$\lambda \mu_{n,p} = \frac{\sum_{j=1}^{n}(1/j)P_{j-1}(\beta)}{P_{n-1}(\beta)}.$$

Differentiating $\lambda \mu_{n,p}$ with β,

$$\lambda \mu_{n,p} \left[\frac{\sum_{j=1}^{n}(1/j)P_{j-1}'(\beta)}{\sum_{j=1}^{n}(1/j)P_{j-1}(\beta)} - \frac{P_{n-1}'(\beta)}{P_{n-1}(\beta)} \right] < 0,$$

because

$$\frac{-P_n'(\beta)}{P_n(\beta)} = \frac{\beta^n/n!}{\sum_{j=0}^{n}(\beta^j/j!)}$$

decreases strictly with n.

6.10 Compared (6.74) with (6.75), for $0 < T < \infty$,

$$\frac{\mu_n F(T)^n}{1 - F(T)^n} \geq \frac{\int_0^T [1 - F(t)^n] dt}{1 - F(T)^n} - T = \frac{\int_0^T t\,dF(t)^n}{1 - F(T)^n},$$

because

$$\mu_n \geq \frac{\int_0^T t\,dF(t)^n}{F(T)^n},$$

whose right-hand side increases with T from 0 to μ_n.

6.11 The numerator of the left-hand side of (6.83) decreases from $\int_0^\infty t\,dF(t)^n$ to 0 and its denominator increases from 0 to 1. So that, the left-hand side decreases from ∞ to 0.

6.12 Differentiating $\int_0^T[1-e^{-\beta\overline{F}(t)}]dt/[1-e^{-\beta\overline{F}(T)}]$ with respect to β,

$$\frac{\overline{F}(T)e^{-\beta\overline{F}(T)}}{[1-e^{-\beta\overline{F}(T)}]^2}\left\{\frac{e^{\beta\overline{F}(T)}-1}{\overline{F}(T)}\int_0^T\overline{F}(t)e^{-\beta\overline{F}(t)}dt-\int_0^T[1-e^{-\beta\overline{F}(t)}]dt\right\}.$$

Letting $L(T)$ be the bracket where $f(t)$ is a density function of $F(t)$,

$$L(0)=0,$$

$$L'(T)=\frac{f(T)e^{\beta\overline{F}(T)}\int_0^T\overline{F}(t)e^{-\beta\overline{F}(t)}dt}{[\overline{F}(T)]^2}[1-\beta\overline{F}(T)-e^{-\beta\overline{F}(T)}]<0,$$

which implies that $L(T)<0$, i.e. $\int_0^T[1-e^{-\beta\overline{F}(t)}]dt/[1-e^{-\beta\overline{F}(T)}]$ decreases with β. Because $\int_0^\infty[1-e^{-\beta\overline{F}(t)}]dt$ increases with β, T_2^* increases strictly with β. Furthermore, we can prove

$$\int_0^\infty[1-e^{-\beta\overline{F}(t)}]dt\geq e^{\beta\overline{F}(T)}\int_0^T[1-e^{-\beta\overline{F}(t)}]dt-T[e^{\beta\overline{F}(T)}-1],$$

because the right-hand side increases with T to $\int_0^\infty[1-e^{-\beta\overline{F}(t)}]dt$.

6.13 Assume that $p_j=\Pr\{n=j\}$ $(j=0,1,2,\ldots)$ in (6.94) and (6.95).

Chapter 7

7.1 Differentiating (7.8) with respect to T_j,

$$-(c_E+c_0)\overline{G}(T_j)-[c_1+(c_E+c_0)(T_{j+1}-T_j)]g(T_j)+(c_E+c_0)\overline{G}(T_{j-1}).$$

7.2 Use a mathematical induction and differentiate $(e^{\theta L}-1)/\theta$ with θ.

7.5 Setting that $1-e^{-\theta t}\equiv x$,

$$\int_0^L[1-(1-e^{-\theta t})^N]dt=\frac{1}{\theta}\sum_{j=0}^{N-1}\int_0^{1-e^{-\theta L}}x^j\,dx=\frac{1}{\theta}\sum_{j=1}^N\frac{(1-e^{-\theta L})^j}{j}.$$

Furthermore,

$$\sum_{j=1}^{N+1} \frac{1}{j} \sum_{j=1}^{N} \frac{(1 - e^{-\theta L})^j}{j} - \sum_{j=1}^{N} \frac{1}{j} \sum_{j=1}^{N+1} \frac{(1 - e^{-\theta L})^j}{j}$$

$$= \sum_{j=1}^{N} \frac{1}{j} [(1 - e^{-\theta L})^j - (1 - e^{-\theta L})^{N+1}] > 0.$$

7.6 In (7.25),

$$\sum_{j=0}^{n-1} \sum_{i=0}^{N-1} \int_0^\infty \frac{(\lambda t)^j}{j!} e^{-\lambda t} \frac{(\theta t)^i}{i!} e^{-\theta t} dt$$

$$= \sum_{j=0}^{n-1} \sum_{i=0}^{N-1} \binom{i+j}{i} \left(\frac{\lambda}{\theta + \lambda}\right)^j \left(\frac{\theta}{\theta + \lambda}\right)^i \frac{1}{(\theta + \lambda)} \frac{1}{(i+j)!} \int_0^\infty x^{i+j} e^{-x} dx$$

$$= \sum_{j=0}^{n-1} \sum_{i=0}^{N-1} \binom{i+j}{i} \left(\frac{\lambda}{\theta + \lambda}\right)^j \left(\frac{\theta}{\theta + \lambda}\right)^i \frac{1}{(\theta + \lambda)}.$$

In (7.30),

$$\sum_{j=0}^{N-1} \int_0^\infty [1 - (1 - e^{-\lambda t})^n] \frac{(\theta t)^j}{j!} e^{-\theta t} dt$$

$$= -\sum_{j=0}^{N-1} \sum_{i=1}^{n} (-1)^i \binom{n}{i} \int_0^\infty \frac{(\theta t)^j}{j!} e^{-(i\lambda + \theta)t} dt$$

$$= -\sum_{i=1}^{n} (-1)^i \binom{n}{i} \frac{1}{i\lambda} \left[1 - \left(\frac{\theta}{i\lambda + \theta}\right)^N\right].$$

7.9 Use

$$\sum_{j=0}^{\infty} \binom{j+n-1}{j} x^j = \frac{1}{(1-x)^n}.$$

Chapter 8

8.1

$$\sum_{k=0}^{\infty}\left\{\overline{G}[(k+1)T][\overline{F}(kT)-\overline{F}((k+1)T)]\right.$$

$$+\int_{kT}^{(k+1)T}\left[\int_{kT}^{u}\mathrm{d}F(t)\right]\mathrm{d}G(u)+\int_{kT}^{(k+1)T}\overline{F}(t)\mathrm{d}G(t)\right\}$$

$$=\sum_{k=0}^{\infty}[\overline{F}(kT)\overline{G}(kT)-\overline{F}((k+1)T)\overline{G}((k+1)T)]=1.$$

8.2

$$\sum_{k=0}^{\infty}\int_{kT}^{(k+1)T}\left[\int_{t}^{(k+1)T}\overline{G}(u)\mathrm{d}u\right]\mathrm{d}F(t)$$

$$=\sum_{k=0}^{\infty}\int_{kT}^{(k+1)T}\left[\int_{kT}^{u}\mathrm{d}F(t)\right]\overline{G}(u)\mathrm{d}u$$

$$=\sum_{k=0}^{\infty}\overline{F}(kT)\int_{kT}^{(k+1)T}\overline{G}(t)\mathrm{d}t-\int_{0}^{\infty}\overline{F}(t)\overline{G}(t)\mathrm{d}t,$$

and

$$\sum_{k=0}^{\infty}k\left[\int_{kT}^{(k+1)T}\overline{G}(t)\mathrm{d}F(t)+\int_{kT}^{(k+1)T}\overline{F}(t)\mathrm{d}G(t)\right]$$

$$=\sum_{k=0}^{\infty}k[\overline{F}(kT)\overline{G}(kT)-\overline{F}((k+1)T)\overline{G}((k+1)T)]=\sum_{k=1}^{\infty}\overline{F}(kT)\overline{G}(kT).$$

8.3 Letting $L(\theta)$ be the left-hand side of (8.3),

$$L(0)\equiv\lim_{\theta\to 0}L(\theta)=\frac{\mathrm{e}^{\lambda T}-1}{\lambda}-T>0,\qquad L(\infty)\equiv\lim_{\theta\to\infty}L(\theta)=0,$$

$$L'(\theta)=\frac{\mathrm{e}^{-\lambda T}}{[\theta(\theta+\lambda)]^2}\left\{\lambda(\theta+\lambda)[1-(1+\theta T)\mathrm{e}^{-(\theta+\lambda)T}]\right.$$

$$\left.+\lambda\theta[1-\mathrm{e}^{-(\theta+\lambda)T}]-(\theta+\lambda)^2(1-\mathrm{e}^{-\lambda T})\right\}.$$

Letting $L_1(T)$ be the bracket,

$$L_1(0) = 0, \quad L_1(\infty) = -\theta^2 < 0,$$
$$L_1'(T) = -(\lambda+\theta)^2 \lambda e^{-\lambda T}[1 - (1+\theta T)e^{-\theta T}] < 0,$$

which implies that $L'(\theta) > 0$, i.e. $L(\theta)$ decreases with θ from $L(0)$ to 0.

8.4 Using the approximation $e^a \approx 1 + a + a^2/2$, (8.3) becomes

$$\frac{\lambda T^2}{2} = \frac{c_T}{c_D}, \quad \text{i.e.,} \quad \widetilde{T} = \sqrt{\frac{2c_T}{\lambda c_D}}.$$

Thus, if c_T becomes 4 times, then $\widetilde{T}$ becomes two times approximately.

8.5

$$\sum_{k=0}^{\infty} \int_{kT}^{(k+1)T} [kH(T) + H(t - kT)]dG(t) = \sum_{k=0}^{\infty} \int_0^T \overline{G}(t + kT)h(t)dt.$$

8.9 See (9.6) of [1, p.151].

8.10 When $F(t) = 1 - e^{-\lambda t}$,

$$\int_0^T t \, df^{(j)}(t) = \frac{j(\lambda T)^j}{j!}e^{-\lambda T} - \sum_{i=j}^{\infty} \frac{(\lambda T)^i}{i!}e^{-\lambda T}.$$

Hence,

$$\sum_{j=1}^{\infty} \left[\frac{j(\lambda T)^j}{j!}e^{-\lambda T} - \sum_{i=j}^{\infty} \frac{(\lambda T)^i}{i!}e^{-\lambda T} \right] [1 - W^{(j)}(x)]$$

$$= \sum_{j=1}^{\infty} \left[W^{(j)}(x) \sum_{i=j}^{\infty} \frac{(\lambda T)^i}{i!}e^{-\lambda T} - \frac{(\lambda T)^j}{j!}e^{-\lambda T} j W^{(j)}(x) \right]$$

$$= \sum_{j=1}^{\infty} \frac{(\lambda T)^j}{j!}e^{-\lambda T} \sum_{i=1}^{j} [W^{(i)}(x) - W^{(j)}(x)].$$

8.14 Suppose that the unit is replaced at time T or at failure N, whichever occurs first. Replacing $p_k(T)$ in (8.72) with $F^{(k)}(T) - F^{(k+1)}(T)$,

$$C_F(T; p) = \frac{c_M \sum_{k=1}^{\infty} \overline{P}_k F^{(k)}(T) + c_R}{\sum_{k=1}^{\infty} p_k \int_0^T [1 - F^{(k)}(t)]dt}.$$

Next, suppose that the unit is replaced at time T or at failure N, whichever occurs last. Similarly, from (8.76),

$$C_L(T; p) = \frac{c_M \sum_{k=1}^{\infty} p_k [k - 1 + \sum_{j=k}^{\infty} F^{(j)}(T)] + c_R}{T + \sum_{k=1}^{\infty} p_k \int_T^{\infty} [1 - F^{(k)}(t)]dt}.$$

Index

© Springer-Verlag London 2014

T. Nakagawa, *Random Maintenance Policies*,

Springer Series in Reliability Engineering, DOI 10.1007/978-1-4471-6575-0

Printed in the United States
By Bookmasters